Practical Home Construction/Carpentry Handbook

TAB BOOKS
Blue Ridge Summit, Pa. 17214

First Edition

First Printing—March 1976
Second Printing—July 1976
Third Printing—May 1977

Hardbound Edition: International Standard Book No. 0-8306-6900-0

Paperbound Edition: International Standard Book No. 0-8306-5900-5

Library of Congress Card Number: 76-1552

Practical Home Construction/Carpentry Handbook

Foreword

If you've ever hammered a nail, sawed a two-by-four, or used a tape measure, you can build your own house. Within the pages of this book you'll find all the information you need to know in order to pick the right wood, prepare the foundation, construct the house right from the ground up, and put on all the finishing touches. You don't even have to concern yourself with design; the final pages in this volume include complete conceptual drawings and building plans for eleven tried-and-proven house ideas that have survived those crucial live-in tests of time and cost!

Economy is the key word. The three books contained within this single volume were prepared independently by Forest Service Engineer L.O. Anderson in conjunction with a special staff at the University of Wisconsin. The aim of the program which culminated in publication of these works was to make it convenient and economical for you to sidestep the problems and expense involved in dealing with others, and do the work yourself!

There's nothing magic about house-building so long as you know exactly what to do at each step of construction and precisely how to do it. In this book Anderson tells all the whats and hows, leaving nothing to your imagination or to chance. And you can use the information in these pages to build any one of a number of unique designs, including conventional structures, tubular houses, and even a breathtaking circular home.

There's more to house-building than carpentry—even though a large portion of the material in this book is devoted to woodworking techniques. These's roofing. There's plumbing. There's ventilation. There's foundation work. Masonry. Room layout. Staircase planning and construction. And, perhaps more important today than ever before, there's insulation. All of these and more are covered in depth. All you have to do is follow the straightforward instructions that will serve as your guide from beginning to end.

And you can choose the house design that most closely resembles your family's own individual life style.

Contents

1
INTRODUCTION 9
SELECTION & USE OF WOOD PRODUCTS 10
  Classification of Woods 11
  Foundations—Sills and Beams (House) 13
  Foundations—Plates and Sleepers (House) 14
  Framing—Joists, Rafters, Headers (House) 15
  Framing—Studs, Plates (House) 16
  Subfloors (House) 16
  Wall Sheathing (House) 18
  Roof Sheathing (House) 20
  Plank Roof Decking (House) 21
  Shingles, Shakes (House) 22
  Exterior Trim (House) 23
  Frames and Sash (House) 23
  Siding (House) 24
  Decking and Outdoor Stepping (House) 26
  Interior Trim With Natural Finish (House) 26
  Interior Trim With Paint Finish (House) 27
  Underlayment for Finish Floors 28
  Flooring—Strip and Wood Block (House) 29
  Miscellaneous Millwork (House) 31
  Paneling (House) 33
  Sills on Foundation Walls (Barn) 35
  Studs, Plates (Barn) 36
  Joists, Rafters (Barn) 37
  Roof Sheathing (Barn) 37
  Siding and Barn Boards (Barn) 38
  Feed Rands and Feed Bunks 40
  Fence Posts 40
  Gates, Fences 41
  Concrete Forms—Framing and Sheathing 42
  Scaffolding 43
  Exposed Platforms and Porches 44
  Tanks, Vats, Storage Bins 44
  Sheathing Papers, Vapor Barriers, and Other Sheet Materials 45
 Selection of Woods According to Important Properties............47
 Surface Characteristics of Common Grades of Lumber 69

Lumber Grades	**71**
Standard Lumber Items	**78**
Other Sheet Materials	**91**
Structural Insulating Board	91
Medium Hardboard	92
High-Density Hardboard	93
Particleboard	93
Interior Finish and Millwork	**93**
Important Points in Construction and Maintenance	**95**
Construction Details	105
Preservative-Treated Materials	111
Classes of Insulation Materials	118
Where to Insulate	119

2 HOW TO CONSTRUCT YOUR LOW-COST HOME ... 121

Major House Parts	**122**
Material Selection	**123**
Joists for 3 Floor Systems	**156**
Framed Wall Systems	**165**
Roof Systems	**179**
Chimneys	**217**
Roof Coverings	**218**
Exterior Wall Coverings	**223**
Framing Details	**233**
Plumbing and Other Utilities	**246**
Thermal Insulation	**251**
Ventilation	**258**
Interior Wall and Ceiling Finish	**264**
Floor Coverings	**276**
Interior Doors, Frames, and Millwork	**284**
Porches	**307**
Steps and Stairs	**317**
Painting and Finishing	**328**
Glossary of Housing Terms	**336**
Foundation Enclosures	**352**

3 DESIGNS FOR LOW-COST WOOD HOMES ... 375

Index ... 436

Introduction

Today it is more important than ever to select the most appropriate wood product for each use in residential and farm construction. Wood products are now being made in more forms and from a greater variety of species than ever before. What was most suitable for a particular use a few years ago may not be so today.

About 25 billion board feet of wood products are used each year by the construction industry in the United States—much of this for homes and farm buildings. In addition, more than 6 billion board feet of lumber are used annually to maintain, repair, and remodel structures.

The wood-based panel products [1] industry produces another 28 billion square feet of material in various thicknesses. Most of this material is used in new construction or remodeling.

This is intended for people who want a reliable source of information for judging and choosing wood products for various purposes. It presents in brief the essential requirements for the usual wood-frame building purposes, and shows how various woods and wood-based products meet these specific requirements. It also emphasizes some basic principles —often overlooked—that should be followed in good construction.

---

[1] Includes wood, particleboard, hardboard, and structural insulating board.

1
SELECTION & USE OF WOOD PRODUCTS

While lumber is widely used in frame construction, sheet materials are also important. Wood-based panel materials are now broadly of three types—plywood, building fiberboard, and particleboard.

Plywood is a glued panel made up of layers of veneer (thin sheets of wood) with the grain of adjacent layers at right angles to each other. The kind of glue used determines whether it is Interior or Exterior type. Plywoods are classified by kinds and by qualities of faces. Those with hardwood faces are usually classed as *decorative* and those with softwood faces as *construction*. Exceptions for softwood plywood include, for example, face veneers of knotty pine or clear, cabinet grades, which are decorative. Plywood is graded on both front and back faces, in that sequence. (For example, A-C, B-B, C-D.)

Building fiberboards are produced with fibers interfelted so the board has some natural bonding. Additives improve the bond and impart strength. Boards of this type are generally classified by density into *structural insulating boards* (with a density of between 10 and about 31 pounds per cubic foot), *medium hardboards* (with a density of between about 31 and 50 pounds per cubic foot), and *high density* or *regular hardboard* (with a density of over 50 pounds per cubic foot).

Particleboards are produced by gluing small particles of wood together into a panel. Hot-setting resins produce the bond necessary to give the panels form, stiffness, and strength. They are generally classified as low density when the board has a density of less than 37 pounds per cubic foot, medium density when the density is between 37 and 50 pounds per cubic foot, and high density when the board weighs more than 50 pounds per cubic foot.

CLASSIFICATION OF WOODS

To select lumber and other wood-based material wisely, one must first single out the key requirements of the job. Then it is relatively easy to check the properties[2] of the different woods to see which ones meet these requirements.

A builder or property owner may believe that he needs a strong wood for the siding of his house or barn when he really requires a wood that takes paint well, is resistant to weathering, and develops little or no warping. Or he may think he needs a wood with high bending strength for the joists of his house, whereas adequate stiffness is more important. Other considerations include the moisture content of the wood, its ability to resist distortion (warping), and its shrinkage characteristics.

In buying sheathing material, one should consider not only the original cost but cost of application as well. Such factors as relative nail-holding qualities, insulation values, and the

---

[2] Distinguishing characteristics, qualities, or marks common to a species or group, usually classified as physical, mechanical, or chemical properties.

possible elimination of corner bracing should also be considered.

It is not necessary to purchase only the best quality lumber or wood-based products. Lower and cheaper grades serve satisfactorily for many uses.

The number of uses and the service requirements of wood vary so greatly that it is practically impossible to classify woods precisely according to their suitability for different uses solely on factual data. Such data, however, can be supplemented by the mature judgment of technical workers who have been impartially studying and testing the various woods for years, and have observed the performance of many woods under widely varying conditions. The opinion of such workers has been included, therefore, in classifying common U.S. wood species for principal home and farm uses.

Wood species are divided into two classes—**hardwoods**, which have broad leaves, and **softwoods** or conifers, which have scalelike leaves or needles. The terms "hardwood" and "softwood" do not denote hardness or softness of the wood. In fact, some "hardwoods" like cottonwood and aspen are less dense (or hard) than some "softwoods" like southern pine and Douglas-fir.

The native species here listed are in general use and are classed conservatively for each specific purpose. Occasionally a species may be underrated for a particular use, or its range of suitability may be underestimated, but the ratings are on the side of safety from the general public's standpoint.

The following classification is simple and applies to average, typical conditions under which wood serves in a particular use. No attempt has been made to draw fine distinctions between woods. Neither is it to be inferred that all species of woods in the same class are equally suitable.

Grades vary considerably by species. Therefore, in this publication, a sequence of first, second, third, fourth, and fifth grade material is given for specific uses. In general, the first grade is for a high or special use, the second for better than average use, the third for average, and fourth and fifth for more economical, but still acceptable construction.

Foundations—Sills and Beams (House)

Usual requirements: High stiffness and strength when used as a beam, good decay resistance, good resistance to withdrawal and lateral movement of nails. Good strength in compression perpendicular to grain (sills). Most woods are satisfactory as sills where dry conditions prevail, but for predominantly wet conditions, preservative-treated wood should be used.

Woods combining usual requirements in a *high* degree: White oak. (Fine for sills and beams in crawl spaces. Heartwood has high decay resistance but wood that is all heartwood usually costs more.)

Douglas-fir, western larch, southern yellow pine, and rock elm. High in strength and nail-holding qualities. (Sills and beams in basement or dry areas. Under moist conditions they require preservative treatment.)

Woods combining usual requirements in a *good* degree: Cedar and redwood. (Sills only, as these species do not have the high bending strength desirable for beams. Heartwood has high decay resistance.)

Poplar, eastern and west coast hemlock, red oak. (Require good preservative treatment if exposed to moist conditions or long periods of high humidity.)

Woods combining usual requirements in a *fair* degree: Ash, beech, birch, soft elm, maple, and sycamore. (Good as beams and fair as sills, but require good preservative treatment if exposed to moist conditions.)

Northern white pine (eastern) and Idaho white pine (western), ponderosa pine, sugar pine, spruce, and white fir. (Satisfactory for sills but require good preservative treatment if exposed to moist conditions or high humidity.)

Grades used: Softwood sills that might be used in houses with crawl spaces are generally of second or third grade softwood Dimension material. For less exacting standards, but nonetheless satisfactory for secondary buildings, fourth grade material may be used. If lumber is not treated, all-heartwood pieces should be selected for sills near ground level and in moist areas where condensed moisture may be absorbed by sills. Hardwood sills are usually first grade Dimension in the best construction and second grade in ordinary construction.

Foundations—Plates and Sleepers (House)

Usual requirements: Good natural decay resistance (or treated with preservative) under moist conditions, good nail-holding qualities, medium density. (Used as plates on concrete slab walls where top of wall is near finish grade, as sleepers on concrete slabs for fastening of finish flooring, and similar uses.)

Woods combining usual requirements in a *high* degree: White oak. (Heartwood has high decay resistance but costs more.)

Woods combining usual requirements to a *good* degree: Redwood, Douglas-fir, western larch, southern yellow pine, rock elm, and other medium-density species for normal conditions. Pressure-treated southern yellow pine, red oak, hemlock, and Douglas-fir for damp conditions.

Woods combining usual requirements to a *fair* degree: Northern and Idaho white pine, ponderosa pine, sugar pine, spruce, white fir, ash, beech, birch, maple, and sycamore for normal conditions.

Grades used: Third grade Dimension lumber of most softwood species under normal conditions.

Framing—Joists, Rafters, Headers (House)

Usual requirements: High stiffness, good bending strength, good nail-holding qualities, freedom from pronounced warp. For this use dryness and size are more important factors than inherent properties of the different woods. Allowable spans vary by species.

Woods combining usual requirements in a *high* degree: Douglas-fir, western larch, and southern yellow pine. (Extensively used.)

Ash, beech, birch, maple, and oak. (Seldom used as they are more difficult to obtain in straight pieces and harder to nail and saw than preceding group.)

Woods combining usual requirements in a *good* degree: Eastern and west coast hemlock, eastern and Sitka spruce, lodgepole pine, and white fir.

Northern and Idaho white pine, ponderosa pine, sugar pine, and redwood. (Seldom used because of adaptability to more exacting uses such as millwork, siding, and finish. Lower strength may be compensated for by using larger joists and rafters.) Poplar. (Seldom used.)

Woods combining usual requirements in a *fair* degree: Elm, gum, sycamore, magnolia, and tupelo. (Seldom used.)

Grades used: Second grade Dimension of most softwood species is used in first-class construction. Third grade is used in a large percentage of lower cost dwellings. The fourth grade is satisfactory for small buildings, but contains more crooked pieces than higher grades. Lumber used in trusses is often first grade and second grade, depending on the type of truss, span, species, and the type of member.

Framing—Studs, Plates (House)

Usual requirements: Moderate stiffness and nail-holding qualities, freedom from pronounced warp, and moderately easy workability (easy to saw and nail).

Woods combining usual requirements in a *high* degree: Douglas-fir, western larch, and southern yellow pine. (Extensively used.)

Woods combining usual requirements in a *good* degree: Eastern and west coast hemlock, spruce, white fir, balsam fir, lodgepole pine, and aspen.

Northern and Idaho white pine, ponderosa pine, sugar pine, and redwood. (Seldom used because of adaptability to more exacting uses as finish.)

Woods combining usual requirements to a *fair* degree: Elm, gum, sycamore, and tupelo. (Seldom used.)

Grades used: Because high bending strength is of secondary importance for studs and plates, grades lower than those commonly used for joists and rafters are satisfactory. Third grade softwood Dimension lumber is satisfactory for most dwellings built to good construction standards. Hardwoods in first and second grade Dimension are used in all types of construction.

Subfloors (House)

Usual requirements:

Lumber.—Requirements are not exacting, but moderate stiffness, medium shrinkage and warp, and ease of working are desired.

Plywood.—Moderate stiffness when finish is strip flooring; high stiffness for wood block or resilient finish flooring. Good nail-holding qualities.

Softwood plywoods for use as subfloors with or without underlayment are classified by density, hence stiffness and strength, into groups. For each grouping a limit for span and loading is established. This is shown on each piece of plywood by a number such as "32/16." The first number indicates maximum span when used as roof sheathing and the second number indicates maximum span when used as subfloor. Here "16" indicates that the maximum span for living area space is framing 16 inches on centers.

Woods combining usual requirements in a *high* degree:

Lumber.—Douglas-fir, western larch, and southern yellow pine. (Commonly used.)

Ash and oak. (Seldom used because of adaptability to more exacting uses.)

Plywood.—Group 1 and 2 softwoods such as: Douglas-fir, southern yellow pine, and western larch.

Woods combining usual requirements in a *good* degree:

Lumber.—Hemlock, ponderosa pine, spruce, lodgepole pine, aspen, balsam fir, and white fir. (Commonly used.)

Northern and Idaho white pine, sugar pine, and poplar. (Seldom used because of adaptability to more exacting uses.)

Beech, birch, elm, hackberry, maple, oak. and tupelo. (Not used extensively, harder to work. Maple, elm, and oak often available locally.)

Plywood.—Group 3 and 4 softwoods such as: Cedar, redwood, Sitka and Engelmann spruce, west coast hemlock, noble fir, and white fir.

Grades and types used (minimum recommended):

Lumber.—Third grade softwood boards are used extensively in better quality houses. In lower cost houses, both third and fourth grades are used. The fourth grade is serviceable and does not entail much waste, but is not as tight as the higher grades. When hardwoods are used, second grade boards are commonly used in the more expensive houses and third grade in the lower cost houses.

Plywood.—Standard interior grade (C-D) under ordinary conditions; in baths, kitchens, or when exposed to weather use Standard grade with exterior glue.

Wall Sheathing (House)

Usual requirements:

Lumber.—Easy working, easy nailing, and moderate shrinkage.

Plywood. — Good nail-holding qualities, workability, and resistance to racking.

Structural insulating board and hardboard. —Good resistance to water, to nailhead pull-through, and to racking if properly attached.

Materials combining usual requirements in a high degree:

Lumber.—Cedar, hemlock, northern and Idaho white pine, ponderosa pine, sugar pine, redwood, aspen, spruce, balsam and white fir, basswood, lodgepole pine, and poplar. (Good racking resistance when applied at 45°, but not adequate when applied horizontally without bracing.)

Plywood.—Douglas-fir, southern pine, and western larch.

Structural insulating board and hardboards. —When applied vertically in 4- by 8-foot or longer sheets with perimeter nailing.

Materials combining usual requirements in a good degree:

Lumber.—Douglas-fir, western larch, and southern yellow pine. (Not as workable as previous lumber group.)

Plywood.—Cedar, redwood, Sitka and Engelmann spruce, west coast hemlock, noble fir, and white fir.

Structural insulating board. — Regular-density structural insulating board (about 18 pounds per cubic foot in density) is furnished in 2- or 4-foot widths and, when applied with long edges horizontal, do not provide necessary resistance to racking forces of wind or earthquake so other bracing must be provided.

The more prevalent way to install insulating board sheathing is in 4-foot widths with long edges vertical. With proper fastening around the perimeter and along interior framing, adequate resistance to racking is provided. Manufacturers' recommendations should be followed for fastening.

Hardboards.—Hardboards are not generally used as wall sheathing but may be used as combined siding-sheathing (see section of "Combined Siding-Sheathing.")

Grades and types used:

Lumber.—Most woods are satisfactory for sheathing, though some woods are less time consuming to work than others. The third grade of Common softwood boards makes a serviceable sheathing when covered with a good building paper. First and second grades provide a tighter coverage but still require coverage with building paper. Fourth and fifth grades may be used as sheathing in low-cost houses, but are not generally available. Both entail some loss in cutting. When a hardwood is used for sheathing, second grade boards are adaptable to more expensive houses, and third grade to the lower cost houses.

Plywood.—Most species of plywood can be used with satisfactory results. For exterior finish such as shingles or shakes, thickness of softer plywoods should be increased to obtain greater nail penetration. Use Standard interior (C-D) under ordinary conditions; use Standard interior with exterior glue if house is in an unusually damp location.

Structural insulating board.—Structural insulating board is furnished in three grades—

regular density, intermediate density, and nail-base. Regular density is manufactured in both the ½ and 25/32 inch thicknesses but the other two grades are only made ½ inch thick.

Intermediate density sheathing is somewhat more dense, hence stronger and stiffer, than regular density. Furnished only in 4- by 8-, or 4- by 9-foot sizes. When properly applied with long edges vertical it satisfies racking requirements while the ½-inch-thick regular density board usually does not. Nail-base sheathing is more dense than intermediate-density and in addition to providing racking resistance has sufficient nail-holding strength to hold some kinds of siding on the wall when special nails are used. Insulating board sheathing must be attached to framing with large-headed (roofing) nails or special staples. These fasteners should have a corrosion-resistant coating.

Roof Sheathing (House)

Usual requirements:

Lumber.—Moderate stiffness, good nail holding, little tendency to warp, ease of working.

Plywood.—Adequate stiffness for span and roof loading. Sheathing grade plywoods are classified into groups by density, hence strength and stiffness. Each grouping sets the distance between supports for proper application and performance. Each sheet is marked with a number such as "32/16" as previously listed under "Subfloors."

Woods combining usual requirements in a *high* degree:

Lumber.—Douglas-fir, western larch, and southern yellow pine. (Commonly used.)

Ash, beech, birch, elm, hackberry, maple, oak, and tupelo. (Not extensively used; harder to work.)

Plywood.—Group 1 and 2 softwoods such as: Douglas-fir, southern yellow pine, and western larch.

Woods combining usual requirements in a *good* degree:

Lumber.—Hemlock, ponderosa pine, spruce, lodgepole pine, aspen, and white and balsam fir. (Commonly used.)

Northern and Idaho white pine, sugar pine, redwood, and poplar. (Seldom used because of adaptability to more exacting uses.)

Plywood.—Group 3 and 4 softwoods such as: Cedar, redwood, Sitka and Engelmann spruce, west coast hemlock, noble fir, and white fir.

Grades and types used:

Lumber.—Third grade Common softwood boards are used extensively in better quality houses. In lower cost houses, both third and fourth grades are used. Fourth grade is serviceable but not as tight as third grade. When hardwoods are used, second grade boards can be used in high-quality houses and third grade in lower cost houses.

Plywood.—Use Standard interior grade (C-D) under ordinary conditions; for unusually damp conditions use Standard interior grade with exterior glue.

Plank Roof Decking (House)

Usual requirements: Moderate stiffness and strength, moderate stability, moderate insulating value. (For short to moderate spans of 2 to approximately 16 feet in length, flat and low-pitch roofs.)

Materials combining usual requirements in a *high* degree: Solid or laminated wood decking (edge matched) of southern yellow pine, Douglas-fir, or other softwood ($1\frac{5}{8}$ to $3\frac{5}{8}$ inches thick).

Materials combining usual requirements in a *good* degree: Structural insulating roof deck.
Grades and types used.

Wood decking.—With solid wood for high-quality houses, first grade; slightly lower class, second grade; standard use (houses and garages), third grade. With laminated wood for high-quality houses, first grade (Select or decorative one face); lower cost houses and other buildings, second grade (Service type).

Structural insulating roof deck.—Specially fabricated products. Types vary by: (a) thicknesses (1½, 2, and 3 inches depending on span and insulation requirements), (b) surface treatment, and (c) vapor barrier needs.

Shingles, Shakes (House)

Usual requirements: High decay resistance, little tendency to curl or check, freedom from splitting in nailing. (Roof and sidewalls.)

Woods combining usual requirements in a *high* degree: Cedar, cypress, and redwood. (Principal shingle woods; heartwood only, edge grain.)

Woods combining usual requirements to a *good* degree: Northern and Idaho white pine, ponderosa pine, and sugar pine. (Handmade shingles or shakes from locally grown timber; for best utility require good preservative treatment.)

White oak. (Handmade shingles or shakes from locally grown timber; require care in nailing.)

Grades used:

Roofs.—In western red cedar, cypress, and redwood, first grade shingles (all-heart, edge-grained clear stock) should be used for the longest life and greatest ultimate economy. Other all-heart but not edge-grained grades, such as second grade in redwood, western red cedar, and cypress, are frequently used to reduce initial cost and for low-cost houses and secondary buildings.

Sidewalls.—Same species as used for roofs. For best construction on single-course sidewalls use first grade (all-heart, edge-grained clear). For double-course sidewalls use third grade for undercourse, and first grade for outer course for best construction, and use second grade outer course to reduce costs.

Exterior Trim (House)

Usual requirements: Medium decay resistance, good painting and weathering characteristics, easy working qualities, and maximum freedom from warp.

Woods combining usual requirements in a *high* degree: Cedar, cypress, and redwood. (Heartwood has natural decay resistance, edge grain preferable for best paint-holding qualities. Most adaptable to natural finishes and stains.)

Northern and Idaho white pine, ponderosa pine, and sugar pine. (Adaptable to ordinary trim.)

Woods combining usual requirements in *good* degree: West coast hemlock, ponderosa pine, spruce, poplar, Douglas-fir, western larch, and southern yellow pine. (Edge-grained boards and special priming treatment advisable to improve paint-holding qualities.)

Grades used: First grades (A, B, or B and Better Finish) are used in the best construction. Second grades (C and D Finish) are used in more economical construction, and first or second grade Common boards where appearance is not important. Clear finger-joint boards are often used when trim is to be painted.

Frames and Sash (House)

Usual requirements: Good to high decay resistance, good paint holding, moderate shrinkage, freedom from warping, good nail holding, and ease of working.

Woods combining usual requirements in a *high* degree: Cypress, cedar, and redwood. Northern and Idaho white pine, ponderosa pine, and sugar pine. (Principal woods used for sash and window and outside doorframes. Usual preservative treatment consists of 3-minute dip in water-repellent preservative.)

Woods combining usual requirements in a *good* degree: Douglas-fir, western larch, and southern yellow pine. (Require dip treatment.)

White oak. (Harder to work and higher shrinkage than softwoods. Usually used for outside doorsills and thresholds.)

Grades used: Grades of lumber for sash and frames are Shop grades and are of primary interest to manufacturers rather than users. The majority of door and window frames and sash are treated with water-repellent preservative at time of manufacture. Decay-resistant species should be considered for basement frames and sash where resistance to moisture and decay is more important. Under severe moisture conditions pressure-treated material is desirable.

Siding (House)

Usual requirements: Good painting characteristics, medium decay resistance, easy working qualities, and freedom from warp. (For lap siding, drop siding, matched vertical boards, vertical boards and battens, etc.)

Lumber siding.—Woods combining usual requirements in a *high* degree: Western red cedar, cypress, and redwood. (Extensively used. Heartwood preferable; edge-grained siding has best paint-holding qualities. Most adaptable of species to natural finishes and stains.)

Woods combining usual requirements in a *good* degree: Northern and Idaho white pine, sugar pine, and white cedar. (Heartwood has medium decay resistance.)

West coast hemlock, ponderosa pine, spruce, and poplar. (Edge grain for best paint retention in such species as hemlock.)

Woods combining usual requirements in a *fair* degree: Douglas-fir, western larch, and southern yellow pine. (Edge grain only.)

Grades used: Redwood and cypress first siding grades (Clear Heart) and western red cedar in a first grade (Clear) for best quality construction. In other softwoods the first grade (B and Better) siding is used in best quality houses. Siding in more economical types of construction is usually of second grade (C or D), but third grade (No. 1 and No. 2) is available in a number of species. Rough-sawn siding patterns in the lower grades are suitable for stain finishes.

Other siding materials.—Paper-overlaid plywood or lumber (resin impregnated in the paper overlay) in sheet form or in manufactured forms for board and batten effect, and in patterns for horizontal siding. Rough-textured plywoods in various patterns and exterior grades are suitable for stain finishes.

Medium hardboard in densities of 32 to 50 pounds per cubic foot. In sheet form or in manufactured widths for horizontal siding. May be plastic coated or factory primed ready for finish paint coats.

High-density or regular hardboards in densities of 50 to 70 pounds per cubic foot. In panel form only. Four- by 8-foot or longer applied with long edges vertical. Such hardboards are not usually recommended for use as lapped (clapboard) siding.

Combined sheathing-siding.—Wood-base panel products can provide both the function of sheathing and siding when applied in large sheets to provide racking resistance and reduction of air infiltration. Special plywoods like rough-sawn western redcedar and "Texture 111" with exterior gluelines are manufactured for this use.

Medium-density hardboards and to a limited extent high-density hardboards are also manufactured for this use. They may have a plain or embossed surface. Plywood and hardboard may be grooved to create reversed board and batten effects, or may have a plain surface and be applied with battens to create the board and batten effect.

Plywood is usually stained; hardboard may be painted or stained.

Decking and Outdoor Stepping (House)

Usual requirements: High decay resistance, nonsplintering, good stiffness, strength, wear resistance, and freedom from warping. (If painted, should have good paint retention.)

Woods combining usual requirements in a *high* degree: White oak. (Edge grain.)

Locust and walnut. (Usually unavailable except when cut from locally grown timber.)

Woods combining usual requirements in a *good* degree: Douglas-fir, western larch, redwood, cedar, and southern yellow pine. (Edge grain only, heartwood preferred.) For moderate life, Douglas-fir and southern yellow pine require preservative treatment. (Softer woods not as wear resistant.)

Grades used: Second (C Finish) or a higher grade in softwoods and first and second Finish grades in hardwoods are used in high-quality construction. In lower cost construction, first grade Dimension in hardwoods and as low as second grade Dimension in softwoods are used. First and second grades in softwoods are serviceable but wear unevenly around knots.

Interior Trim with Natural Finish (House)

Usual requirements: Hardness, freedom from warp, pleasing texture and grain.

Woods combining usual requirements in a *high* degree: Oak, birch, maple, cherry, beech, sycamore, and walnut. Cypress (pecky) and maple (curly or bird's eye). *Knotty surface.—*Cedar, ponderosa pine, spruce, sugar pine, gum, and lodgepole pine.

Woods combining usual requirements in a *good* degree: Douglas-fir, west coast hemlock, western larch, southern yellow pine, redwood, aspen, and magnolia. (With conventional architectural treatment.)

Grades used: High-class hardwood interior trim is usually first grade Finish (A grade). The softwood Finish Grade A or B and Better is commonly used in high-quality construction. In the more economical types of construction, C grade is serviceable. D grade requires special selection or some cutting to obtain clear material. Special grades of knotty pine, pecky cypress, and others are available to meet special architectural requirements in some types of high quality construction.

Interior Trim With Paint Finish (House)

Usual requirements: Fine and uniform texture, moderate hardness, absence of knots and discoloring pitch, good paint holding, and freedom from warp and shrinkage.

Woods combining usual requirements in a *high* degree: Northern and Idaho white pine, ponderosa pine, sugar pine, and poplar. (Where likelihood of marring is negligible.)

Woods combining usual requirements in a *good* degree: Hemlock, redwood, spruce, white fir, magnolia, basswood, beech, gum, maple, and tupelo.

Douglas-fir, western larch, and southern yellow pine. (Edge grain most satisfactory.)

Grades used: C Finish is the lowest softwood grade commonly used for high-quality paint and enamel finish. D Finish can be used but requires some selection or cutting. First grade Common is used for ordinary or rough-paint finishes. In more economical homes second grade Common may be used for ordinary or rough-paint finishes. Smooth-paint finishes are difficult to obtain and maintain over knots in first, second, and third grade Common softwoods.

First grade Finish in the hardwoods is used for exacting requirements of high-quality paint and enamel finish in more expensive homes. The second grade Finish in hardwoods is also used but requires some selection or cutting. Second grade boards in hardwoods may be used for interior trim in the low-cost home, but for interior trim that is to be painted softwoods are generally used.

Underlayment for Finish Floors

Ordinarily all finish flooring except standard strip flooring and ½- or ¾-inch wood block floor are laid with an underlayment between the subfloor and the finish flooring. This is especially necessary for resilient floor surfacing (rubber, vinyl, vinyl asbestos, or asphalt in tile or sheet form) because of its thinness, flexibility, and tendency to "showthrough" the pattern of the surface beneath it.

Floor underlayment serves the following functions:

1. Provides uniform support for finish flooring.
2. Bridges small irregularities in the subsurface.
3. Because joints in floor underlayment do not coincide with those in subfloor, there is less chance for working of joints to loosen or break finish flooring.
4. Provides a smooth, uncontaminated surface for gluing to the base those kinds of finish flooring requiring it.
5. Permits vertical adjustment in floor levels so all rooms are at the same elevation even when different floorings are used. The subfloor usually serves as the working platform. During the period between initial laying of subfloor and installation of finish flooring the surface may be roughened from wetting, dented from impacts, or contaminated with plaster, dirt,

grease, and paint, in fact anything that is tracked or brought into the building.

Some use of combined subfloor-underlayment is developing, particularly in factory-built or tract-built housing where subfloors are given special protection during construction or where pad and carpet are installed.

Floor underlayments are plywood, hardboard, or particleboard.

Plywood underlayment.—Plywood underlayment is a special grade produced for this purpose from group 1 woods (for indentation resistance). It is produced in ¼-, ⅜-, ½-, ⅝-, and ¾-inch thickness, and the face ply is C plugged grade (no voids) with a special C or Better veneer underlying the face ply to prevent penetration from such concentrated loads as high heels.

Particleboard underlayment.—Produced in the same thicknesses as plywood, particleboard underlayment is often preferred because its uniform surface and somewhat higher density make it more resistant to indentation than plywood when thin resilient flooring is applied over it. Because it tends to change more in length and width with changes in moisture content than plywood, manufacturers' directions for installation and specifications for adhesives must be followed for good performance.

Hardboard underlayment.—Produced in 4-foot squares, 0.220 inch thick and planed to uniform thickness, hardboard underlayment should be installed to manufacturers' specifications for proper performance. It is mainly used on remodeling or in new construction where minimum thickness buildup is desired.

Flooring—Strip and Wood Block (House)

Usual requirements: High resistance to wear, attractive figure or color, minimum warp and shrinkage. (Material should be used at a moisture content near the level it will average in service.)

Woods combining usual requirements in a *high* degree: Maple, red and white oak, beech, and birch. (Most commonly used hardwoods.)

White ash and walnut. (Not commonly used.)

Hickory and pecan. (Not commonly available. Harder to work and nail. More suitable to woodblock flooring.)

Woods combining usual requirements in a *good* degree: Cherry, gum, and sycamore (edge grain). (Not commonly available. Highly decorative and suitable where wear is not severe.)

Cypress, Douglas-fir, west coast hemlock, western larch, and southern yellow pine (edge grain). (More suitable in low-cost houses in bedrooms where traffic is light.)

Grades used: In beech, birch, and maple flooring the grade of Firsts is ordinarily used for better quality homes, and Seconds and sometimes Thirds in economy houses. In oak, the grade of Clear (either flat or edge grain) is used in better construction, and Selects and sometimes No. 1 Common in lower cost work or where small tight knots provide the desired effect. Other hardwoods are ordinarily used in the same grades as oak.

When softwood flooring is used (without covering) in better quality homes, grade A or B and Better edge grain is used. Grade D or C (edge grain) is used in low-cost homes.

The three general types of material used for finish floors are wood strip, wood block, and resilient flooring such as rubber, vinyl, or asphalt tile, or linoleum.

Strip flooring.—Is usually laid over boards nominally 1 inch thick because the boards must be thick enough to hold the nail. For best results, boards for subfloors are laid diagonally and in nominal widths no greater than 6 or 8 inches. Plywood ⅝ or ¾ inch thick is also satisfactory. One-half inch plywood is satisfactory for subfloor when strip flooring is nailed to floor joists.

Wood block flooring.—Because wood block flooring requires an even and uniform base for best results, plywood subfloor is frequently used. A ⅝- or ¾-inch thickness should be used if block flooring is installed by nailing. Laminated block flooring ½ inch thick or less may be used over a ¼- or ⅜-inch plywood or particleboard underlayment that has been nailed to a wood subfloor.

Resilient flooring.—Because resilient floors are usually quite thin and are installed with adhesives, it is necessary to provide a smooth base. Plywood, particleboard, or hardboard—all of a special underlayment grade—are most frequently used over various types of subfloors. Underlayment screws are commonly used to fasten the underlayment and minimize "popups" that can occur with other fastenings.

Miscellaneous Millwork (House)

Interior millwork usually varies a great deal between houses, both in the type and amount used. Uses in the average homes include doors, kitchen cabinets, shelving, and stairs. Other homes may, in addition, involve the use of such millwork items as fireplace mantels, wall paneling, ceiling beams, china closets, bookcases, and wardrobes.

Doors

Usual requirements: Freedom from warp (especially for outside doors), good finishing qualities, resistance to denting (hardness), pleasing figure or grain for natural finish or good base for paint.

Other attributes and sometimes requirements of doors include resistance to fire and sound transmission, ability to hold special hardware, means to accept cutouts or openings for windows, and durability. Doors use either an interior or exterior quality glue for assembly, depending on where they are to be used.

Door manufacture is of two types—the panel door with insert panel and solid or veneered stiles and rails, and the flush door with skins bonded to frames. The flush door is manufactured in hollow core construction (for interior doors) and solid core (for exterior doors in cold and moderate climates).

Woods combining usual requirements in a *high* degree: Oak and birch. (Natural finish.)

Woods combining usual requirements in a *good* degree: Ponderosa pine, Douglas-fir, southern yellow pine, and spruce. Gum for natural finish or painting.

Stairways

Usual requirements for treads, risers, and stair parts: Hardness and wear resistance (treads, railings), freedom from warp, pleasing grain.

Woods combining usual requirements in a *high* degree: Oak, birch, maple, walnut, beech, ash, and cherry (exposed treads and risers.)

Woods combining usual requirements in a *good* degree: Douglas-fir, southern yellow pine, gum, and sycamore (basement or secondary stairs or when stairs are to be carpeted).

Cabinet Doors

Usual requirements: Pleasing grain, freedom from warp, moderate hardness.

Woods combining usual requirements in a *high* degree: Maple, oak, birch, and cherry. (Suitable for natural finishes and for plywood flush doors.)

Woods combining usual requirements in a *good* degree: Douglas-fir, southern yellow pine, gum, ponderosa pine, magnolia, and poplar for paint finish.

Shelving

Usual requirements: Stiffness, freedom from warp.

Woods combining usual requirements in a *high* degree: Ash, birch, maple, oak, and walnut. (Suitable for natural finishes.)

Douglas-fir, poplar, southern yellow pine, redwood, ponderosa pine, sugar pine, and Idaho white pine. (Suitable for paint finish.)

Woods combining usual requirements in a *good* degree:

Lumber.—Hemlock, spruce, and western larch.

Plywood.—Natural finish: Oak and birch. (Most available species.) Painted finish: Douglas-fir, southern yellow pine, and other softwoods.

Particleboard.—Though only one-fourth to one-eighth as stiff as wood or plywood, particleboard is being used increasingly where loading is light, extra support is provided, or where spans are short. Frequently veneered or overlaid with higher stiffness materials to provide additional stiffness.

Paneling (House)

Usual requirements (for natural finish or light staining): Pleasing grain, figure or surface treatment, freedom from warp and shrinkage and some resistance to abrasion.

Woods combining usual requirements in a *high* degree:

Lumber.—Oak, redwood, cypress (pecky), walnut, cedar (knotty), ash, birch, pine (knotty), and cherry.

Plywood.—Oak, birch, maple, pecan-hickory, and walnut.

Woods combining usual requirements in a *good* degree:

Lumber.—Gum, western larch, Douglas-fir, beech, southern yellow pine, hemlock, and ponderosa pine.

Plywood.—Cedar, pine, Douglas-fir, southern yellow pine, and some imported species. (Some are specially treated to create a variation in the grain for unique surface effects.)

Grades and types used:

Lumber.—The best grade in hardwood for high quality houses is first grade. Softwood first or second grades are commonly used in the better house. Third grade is more economical. Special grades of knotty pine, pecky cypress, and sound wormy oak are sometimes available for special paneling treatment.

Plywood.—Unfinished: good or special surface one side, interior or exterior types. Prefinished: V-grooved and others (good one side or equal.)

Other materials.—Hardboards with special grain printing, embossing, or other surface treatments or decorative laminate overlays. Structural insulating board in sheet or plank form for walls and in tile form or lay-in panel for ceiling. (Factory treated, finished, or special acoustical effect.) Particleboards with veneered plastic, or other overlay face.

Barns

Wood and wood-based materials for barns and similar buildings are generally the same as those outlined for houses. Grades are usually lower but in some uses strength is the most important factor, and the need for the additional strength is often reflected in the recommended grades and species. Lower grades can be used for siding, flooring, and trim than are ordinarily used for houses.

Poles

Usual requirements: (For load-bearing poles and posts used in such construction as pole barns, the butt end of the pole is usually embedded in the soil.) High stiffness and strength, freedom from crook, minimum taper, good nail holding, good decay resistance. (All poles used in permanent construction should be pressure treated in compliance with Federal Specification TT-W-571. This specification is available from the nearest General Services Administration Business Service Center for 10 cents.)

Woods combining usual requirements in a *high* degree: Western larch, Douglas-fir, west coast hemlock, and southern yellow pine. (Must be preservative treated.)

Woods combining usual requirements in a *good* degree: Lodgepole pine, jack pine, red pine, and ponderosa pine. (Must be preservative treated.)

Woods combining usual requirements in a *fair* degree: Western red cedar and northern white cedar. (Heartwood of cedars has good decay resistance.)

Classes used: The class of pole required in a building is usually determined by the circumference at the top. Classes vary from Class 1 poles with 27-inch minimum top circumference to Class 10 poles with 12-inch minimum top circumference. Poles should bear identification markings. Lengths vary and include distance above ground as well as embedment depth.

Sills on Foundation Walls (Barn)

Usual requirements: Good nail holding, moderate hardness, and good decay resistance (pressure treated in accordance with recognized standard such as Federal Specification TT-W-571 for permanent use). High stiffness and strength are important when piers or posts are used instead of walls, and the sill acts as a beam.

Woods combining usual requirements in a *high* degree: Cedar, redwood, and white oak. (Heartwood has high decay resistance. Cedar and redwood have lower bending strength and are usually not used as beams.)

Woods combining usual requirements in a *good* degree: Douglas-fir, western larch, southern yellow pine, rock elm, and poplar. (High in strength and nail holding.)

Woods combining usual requirements in a *fair* degree: Eastern and west coast hemlock, northern and Idaho white pine, ponderosa pine, sugar pine, spruce, white fir, ash, beech, birch, soft elm, maple, red oak, and sycamore.

Grades used: Softwood sills in large barns are generally of second Dimension grade. Third grade is used in small and low-cost barns. Second and third Dimension grades usually have a high percentage of heartwood. Unless material is pressure treated, all-heartwood pieces should be selected for sills, especially where foundation walls are close to the ground. Hardwood sills are usually of the first Dimension grade in large barns and of the second grade in small barns.

Studs, Plates (Barn)

Usual requirements: Medium stiffness and strength, good nail holding, medium freedom from warp, moderate ease of working. In some barns, especially dairy, preservative treatment or good natural decay resistance is an added requirement. Studs in cribs or granaries are subjected to heavy lateral pressures from stored grain and require strength and stiffness in addition to good fastenings.

Woods combining usual requirements in a *high* degree: Douglas-fir, western larch, and southern yellow pine.

Redwood. (Heartwood decay resistance is high, but has lower bending strength.)

Woods combining usual requirements in a *good* degree: Hemlock, northern and Idaho white pine, ponderosa pine, sugar pine, lodgepole pine, Sitka spruce, white fir, eastern spruce, balsam fir, aspen, and poplar.

Ash, beech, birch, locust, maple, and oak. (Harder to nail and fabricate.)

Elm, gum, hackberry, and sycamore. (More difficult to fabricate and not widely available.)

Grades used: Third Dimension grade is the principal softwood grade used for studs in normal construction, but second grade should be used for cribs and granaries. Fourth grade is serviceable but is more difficult to fabricate because it contains more crooked pieces and entails some loss in cutting. Fourth grade is used in small, inexpensive barns. Hardwoods in second grade Dimension are used in most types of construction.

Joists, Rafters (Barn)

Usual requirements: High stiffness and strength, good nail holding, and moderate ease of working. Woods of moderate bending strength can be used with satisfactory results if lower strength is compensated for by the use of larger members, by closer spacing, or by shorter spans.

Woods combining usual requirements in a *high* degree: Douglas-fir, western larch, and southern yellow pine.

Ash, beech, birch, maple, and oak. (Harder to nail and work, not widely available.)

Woods combining usual requirements in a *good* degree: Hemlock, redwood, Sitka spruce, white spruce, white fir, elm, gum, hackberry, sycamore, tupelo, and poplar.

Woods combining usual requirements in a *fair* degree: Cedar, northern and Idaho white pine, ponderosa pine, sugar pine. Engelmann spruce, aspen, basswood, and cottonwood.

Grades used: The third Dimension grade of most softwood species is used in normal construction. Added strength in large, high-class barns can be obtained by the use of second grade Dimension. For long spans, stress grades might also be considered to eliminate larger sizes or closer spacing of members. The fourth grade of all softwood species is used in small and low-cost barns. The hardwood grades used are second grade for most farm buildings.

Roof Sheathing (Barn)

Usual requirements:

Lumber.—Medium stiffness, good nail holding, low shrinkage, medium decay resistance, freedom from splitting.

Plywood.—Good stiffness, good nail holding, resistance to delamination in high humidities.

Woods combining usual requirements in a *high* degree:

Lumber.—Douglas-fir, western larch, and southern yellow pine.

Plywood.—Group 1 and 2 such as: Douglas-fir and southern yellow pine. (Most available.)

Woods combining usual requirements in a *good* degree:

Lumber.—Hemlock, aspen, lodgepole pine, northern and Idaho white pine, ponderosa pine, sugar pine, spruce, white fir, and redwood. (Render good service in barns with low decay hazard.)

Elm, gum, oak, poplar, beech, birch, and maple. (Sometimes available from locally grown timber.)

Plywood.—Group 3 and 4 such as: Cedar, redwood, Sitka and Engelmann spruce, west coast hemlock, noble fir, and white fir.

Grades and types used:

Lumber.—The third Common board grade is normally used in construction of most barns. The fourth grade is serviceable and may be used in small low-cost barns but usually entails some waste in cutting.

Plywood.—Use Standard interior grade (C-D) for dry conditions and Standard interior grade with exterior glue for damp conditions.

Siding and Barn Boards (Barn)

Usual requirements:

Lumber.—Good painting or weathering qualities, freedom from warping or splitting, medium decay resistance. Medium bending strength when used without sheathing backing or with only a nominal number of cross supports. Boards subjected to dampness from the ground or to constant wetting should have high decay resistance or have a preservative treatment.

Plywood.—Good finishing or weathering qualities, freedom from warping and delamination, medium decay resistance. Medium bending strength in walls with wide spacing of frame members and without interior lining.

Woods combining usual requirements in a *high* degree: Cypress, cedar, and redwood. (Heartwood.)

Woods combining usual requirements in a *good* degree:

Lumber.—Northern and Idaho white pine, ponderosa pine, sugar pine, and poplar. (Heartwood preferable.)

Douglas-fir, western larch, and southern yellow pine. (Edge grain preferable. Should be given special priming coats and protected against weathering by good paint maintenance.)

Plywood.—Group 1 and 2 such as: Douglas-fir and southern yellow pine. (Most available.)

Woods combining usual requirements in a *fair* degree:

Lumber.—Hemlock, spruce, aspen, lodgepole pine, balsam fir, and white fir.

Plywood.—Group 3 and 4 such as: Cedar, redwood, Sitka and Engelmann spruce, west coast hemlock, and white fir. (Also have good paint retention.)

Grades and types used.

Lumber.—The grade of bevel siding is generally higher than the grade used with drop siding or barn boards. When bevel siding is used, it is usually in third and fourth grade, depending on species. When drop siding is used, it is usually second and third grade in the better quality barns. However, the lowest grades are generally serviceable in the lower cost barns. Fourth grade Common boards are also used extensively in lower cost barns. Second grade Common boards are used in the higher quality but entail some loss.

Plywood.—If the building is to be painted, overlaid plywood is used for highest quality buildings. For stained finishes use C-C Exterior (unsanded) grade. Exterior plywood with various surface treatments would also be satisfactory for buildings that are to be stained.

Feed Racks and Feed Bunks

Usual requirements: Hardness, nonsplintering. (Edge grain for most satisfactory outdoor use).

Woods combining usual requirements in a *high* degree: Ash, beech, birch, locust, rock elm, hickory, maple, oak, and soft elm.

Woods combining usual requirements in a *good* degree: Douglas-fir, western larch, southern yellow pine, redwood, gum, and tupelo.

Grades used: The hardwoods are used in first and second Dimension grades, the softwoods in second and third grades. In the more economical type of work, softwood grades as low as fourth grade prove satisfactory.

Fence Posts

Usual requirements: High decay resistance and little or no sapwood for untreated posts. Good bending strength, straightness, and high staple holding. Permanent installation requires a good preservative treatment. High sapwood content is desirable for fence posts to be preservative treated.

Woods combining usual requirements in a *high* degree: Black locust and osage orange. (Meet most requirements, but not readily available in all parts of the United States.)

White oak. (Heartwood only. Generally available in the Eastern States, but life shorter than preceding group if not treated.)

Cedar, cypress, and redwood. (Heartwood only. Readily available but do not hold smooth shank staples and nails so well as preceding groups.)

Woods combining usual requirements in a *good* degree: Douglas-fir, western larch, and southern yellow pine (preservative treatment required).

Woods combining usual requirements in a *fair* degree: Beech, birch, maple, red oak, and elm. (Equal the best woods when given a good preservative treatment.) Hemlock, spruce, white fir, basswood, cottonwood, gum, tupelo, poplar, and lodgepole pine.

Grades used: Fence posts have no standard grades, but are specified by top diameters and by lengths. Treated posts should be branded or stamped to identify the treatment and source.

Gates, Fences

Usual requirements: Good bending strength, good decay and weather resistance, high nail holding, freedom from warp. Treatment desirable for severe conditions. (Should also be lightweight for gates.)

Woods combining usual requirements in a *high* degree: Douglas-fir, western larch, southern yellow pine, redwood, and white oak.

Woods combining usual requirements in a *good* degree: Cedar, northern and Idaho white pine, ponderosa pine, sugar pine, and poplar. (Small tendency to warp, weather well, but are low in strength and nail holding. All except cedar have moderately low resistance to decay.)

Beech, birch, gum, maple, red oak, and tupelo. (Strong, high in nail holding, but have greater tendency to warp, do not weather so well as preceding group, and are too heavy for gates. All except gum and maple have moderately low resistance to decay.)

Eastern and west coast hemlock, white fir, and spruce. (Intermediate qualities except for decay resistance, which is moderately low.)

Grades used: Second and third grade softwood Common boards and second hardwood board grades are used in better and more substantial gates and fences. In smaller and more economical gates and fences, third grade hardwood boards are used. A softwood grade as low as fourth grade Common boards may be used, but entails some loss because of cutting out the larger defects.

Concrete Forms—Framing and Sheathing

Usual requirements:

Lumber framing and sheathing.—Good stiffness, good bending strength, resistance to warping and splitting during installation and reuse, ease of working, smooth surface.

Plywood sheathing.—Good stiffness, water resistance, good bending strength, resistance to warping.

Woods combining usual requirements in a *high* degree:

Lumber framing and sheathing.—Douglas-fir, western larch, and southern yellow pine—often with fiber overlays when forms are reused. (High strength, good reuse value.)

Plywood sheathing.—Douglas-fir, western larch, and southern yellow pine.

Woods combining usual requirements in a *good* degree:

Lumber framing and sheathing.—Northern and Idaho white pine, eastern and west coast hemlock, ponderosa pine, redwood. white fir, white spruce, and Sitka spruce. (Easy to cut and nail.)

Plywood sheathing.—Cedar, redwood, Sitka and Engelmann spruce, west coast hemlock, noble fir, and white fir. (Almost any species of plywood is suitable for formwork with correct spacing of supports.)

Grades and types used:

Lumber.—With compensation in size of material or in frequency of bracing, almost all woods can be used in ordinary construction for concrete forms. Second and third Common board grades of softwoods in dressed and matched or shiplap form, and second board grade hardwoods are used for coverage in forms with minimum bracing. Fourth grade softwoods or third grade hardwoods are used for forms in which the spacing is close or the loads are small. Framing and bracing utilizes third or fourth grade Dimension in softwoods and second grade Dimension in hardwoods. Concrete forms lined with "formboard" hardboard permit the use of lower grade boards than if lumber alone is used.

Plywood.—"Plyform" with sanded faces and mill oiled is most commonly used.

Scaffolding

Usual requirements: High bending strength, high stiffness, high nail holding, medium weight, and freedom from compression failures and crossgrain.

Woods combining usual requirements in a *high* degree: Douglas-fir, western larch, and southern yellow pine.

Woods combining usual requirements in a *good* degree: Redwood, spruce, and west coast hemlock. (Lower bending strength.) Birch, white ash, elm, maple, and oak. (Harder to saw and nail.)

Woods combining usual requirements in a *fair* degree: Sugar pine, ponderosa pine, and Idaho white pine. (Low stiffness and strength.)

Grades used: First grade softwood Dimension is usually required for scaffolding that must support loads under conditions that involve hazards. Light scaffolding may be selected from second grade softwood Dimension; in hardwoods, uprights can be selected from first grade Dimension. Selection should eliminate all pieces with compression failures, large or unsound knots, and crossgrain.

Some State building codes designate the grades to be used for scaffolding. Southern pine and western grading rules include special scaffolding plank grades.

Exposed Platforms and Porches

Usual requirements: High decay resistance, good stiffness and strength for framing, and good wear and splinter resistance for decking. (Where wood is exposed to severe moisture conditions, treated material is recommended.)

Woods combining usual requirements in a *high* degree: Redwood, locust, and white oak. (Heartwood only.)

Woods combining usual requirements in a *good* degree: Cedar, Douglas-fir, western larch, southern yellow pine, and rock elm. (Edge grain.)

Grades used: First or second grade Dimension in softwoods and first grade Dimension in hardwoods are the grades ordinarily used.

Tanks, Vats, Storage Bins

Usual requirements: High decay resistance and low shrinkage. Treated wood should be used in storage of silage at or below grade.

Woods combining usual requirements in a *high* degree:

Lumber.—Cedar, cypress, and redwood. (When untreated, heartwood only.)

White oak. (When untreated, edge-grained heartwood only.)

Woods combining usual requirements in a *good* degree:

Lumber.—Douglas-fir, western larch, and southern yellow pine. (Treated edge-grained material preferred.)

Plywood.—Group 1 woods if preservative treated.

Woods combining usual requirements in a *fair* degree:

Lumber.—Beech, birch, eastern spruce, hemlock, northern and Idaho white pine, and ponderosa pine.

Plywood.—Cedar, redwood, Sitka and Engelmann spruce, west coast hemlock, noble fir, and white fir. (All good paint retention.)

Other Group 2, 3, and 4 woods.

Grades and types used:

Lumber.—The requirements for silos, tanks, and vats are best met by grades prepared especially for these uses. Such special grades are sold as tank, tank and boat, or silo stock, and are available in most of the softwoods well adapted to these uses. The clear-heart grades available in cypress and redwood also are used extensively where requirements are high. There are no special grades in hardwoods for silos, tanks, or vats. Hardwoods, when used, should be bought on special order calling for all-heart, tight stock.

Plywood.—If the silo or other structure is to be painted, use paper-overlaid plywood. If the structure is to be stained, use Exterior, B-C, for higher quality structures, Exterior, C-C plugged, for lower cost buildings.

Sheathing Papers, Vapor Barriers, and Other Sheet Materials

Sheathing paper and vapor barriers have several general uses in the construction of houses and other frame buildings. For example, sheathing paper resists moisture and wind infiltration when used over unsheathed walls, over lumber sheathing, over all types of sheathing materials with a stucco exterior finish, and as backing for masonry veneer. The paper for such purposes should be waterproof but of the "breathing" type. This allows any escaped water vapor to move through the paper and minimizes condensation problems. Many types of materials are available for this use, including 15-pound asphalt-saturated felt.

Paper (15- or 30-pound felt) is also used as a roof underlayment for asphalt shingles when roof slopes are less than 7 in 12. Such protection is usually not needed under wood shingles except as an eave flashing to prevent moisture entry from ice dams.

Roll roofing in 45-pound and heavier weights may be used for roofing small buildings and temporary structures. Built-up roofing, consisting of a number of plies of 15- and 30-pound asphalt-saturated felt, is used on low-pitch or flat roofs. For wood decks, a nailed sheet is placed over the deck before installing alternate layers of felt and asphalt or pitch. This type of roof is usually topped with gravel or crushed stone.

Paper or deadening felt is often desirable under finish floor, as it will stop a certain amount of dust and deaden the transfer of sound.

Vapor barriers are used in walls, floors, and ceilings, usually in conjunction with insulation, to minimize the movement of water vapor to cold, exposed surfaces. They are effective in livestock barns as well as dwellings. Vapor barriers consist of plastic films, laminated or coated papers, or aluminum foil. For protection from cold weather condensation, they should be applied as close to the inner warm surface as possible, usually just under the interior coverings. They are also used under concrete slabs to prevent ground moisture from coming through.

Vapor barriers are also used as ground covers in crawl spaces to prevent wood framing and other wood materials from becoming damp from ground moisture. These barrier materials consist of duplex paper with asphalt laminate, plastic films, aluminum foil backed with paper, roll roofing, and various combinations of materials.

Product Standards

A number of quality standards have been developed to insure the quality of wood-based products used in the construction of houses and other wood-frame construction. For example, a quality control policy covers the fabrication of plywood components, and an inspection manual is available for structural glued-laminated wood members. Similar standards, such as manufacturers' procedures, provide some control in the production of various manufactured items.

Another system of controlling quality in various materials and fabricated units is through National Product Standards (formerly Commercial Standards). These standards, published by the U.S. Department of Commerce, establish quality levels for manufactured products in accordance with the principal demands of the trade. They give technical requirements for materials, construction, dimensions, tolerances, testing, and other details to promote sound commercial practices in the manufacture, marketing, and application of the products. The standards are developed by voluntary cooperation among manufacturers, distributors, consumers, and other interests.

Many Product Standards relate to wood and wood-based materials.

SELECTION OF WOODS ACCORDING TO
IMPORTANT PROPERTIES

The choice of one wood species in preference to another for any of the principal home or farm uses should seldom be based on a single vital property. Usually a favorable combination of two or more basic qualities or characteristics should determine the selection.

In table 1 the various woods are classified according to a number of these important properties. This table is helpful when a wood is not listed for a specific use or if it departs markedly from the general classification previously outlined. Class A, in table 1, includes woods that are relatively high in the specific property or characteristic listed; class B woods are intermediate in the specific property or characteristic listed; and class C woods are relatively low.

For example, Class A in columns 2 to 16 indicates species that are generally the most desirable from the standpoint of working and behavior characteristics and strength. In columns 17 to 21, class A indicates species with the most

47

desirable qualities because it designates greater freedom from knots and other characteristics, and greater acceptability as to their size.

Such a general classification necessarily ignores small differences and sacrifices detail in favor of the simplicity desired by the ordinary user. All woods in the same class are by no means equal, and no attempt is made to draw fine distinctions between the species.

For the different kinds (species) of wood, table 1 assumes equal size, equal dryness, and, for strength properties, an equal number of knots and other strength-reducing characteristics. So far as cross-sectional dimensions are concerned, in actual practice the different species of softwood lumber are all governed by trade standards.

Standard sizes for boards and dimension are larger for lumber surfaced green than lumber surfaced dry. When lumber surfaced green dries to the standard dry moisture content, it will shrink to approximately the standard dry surfaced size.

Most hardwoods differ substantially from softwoods in their properties (basic characteristics) and in their uses. As a class, hardwoods are heavier, harder, shrink more, and are tougher. Hardwoods and softwoods are similar in stiffness, so on a weight basis the softwoods are actually much stiffer. In strength as a post and in bending strength the two groups are more directly comparable than they are in weight, toughness, and hardness; nevertheless, more commercial hardwoods than softwoods can be rated high in bending strength.

The softwoods are used principally in construction work, whereas hardwoods furnish most of the wood for interior finish and flooring as well as for implements, furniture, and other industrial uses. In addition to normal construction uses. 2-inch and thicker lumber is also sold stress-graded for more carefully engineered components such as trusses.

The various properties and characteristics as designated in table 1 are more fully described in the following sections.

Hardness

Hardness (table 1, column 2) is the property that makes a surface difficult to dent, scratch, or cut. Generally, the harder the wood, the better it resists wear, the less it crushes or mashes under loads, and the better it can be polished. On the other hand, the harder wood is more difficult to cut with tools, harder to nail, and more likely to split in nailing.

Hardness is of particular concern in flooring, furniture, and tool handles. Hardness is also important in selecting interior trim such as door casings, base, and base shoe, as well as door jambs, sills, and thresholds. These portions usually receive the hardest wear in a house.

There is a pronounced difference in hardness between the springwood and the summerwood of woods such as southern yellow pine and Douglas-fir. In these woods the summerwood is the denser, darker colored portion of the annual growth ring. Differences in surface hardness thus occur at close intervals on a piece of such wood depending on whether springwood or summerwood is encountered. In woods like maple, which do not have pronounced springwood and summerwood, the hardness of the surface is quite uniform.

The classification of a species as a hardwood or softwood is not based on actual hardness of wood. Technically, softwoods are those cut from coniferous or evergreen trees, whereas hardwoods are those cut from broad-leaved and deciduous trees. Actually, some of the softwoods are harder than some of the hardwoods.

As a group, the hardwoods can be divided into (a) dense and (b) less dense. The softwoods can also be divided into two groups: (a) medium-density and (b) low-density.

A number of woods are strong favorites for building purposes largely because of their softness and uniformity rather than their hardness. Northern white pine (eastern) and Idaho white pine (western), poplar, white fir, and basswood are traditional examples. Others are ponderosa pine, sugar pine, and cedar. The ease with which these woods can be cut, sawed, and nailed has put them in a high position for general use. This is less important in present-day construction because portable power tools make it easier to handle such dense species as Douglas-fir and southern yellow pine. In fact, the use of these denser species allows greater spans for joists and rafters than can be used for equal-sized members of the softer woods.

Differences in hardness are great enough to affect the choice of woods for such uses as flooring and furniture on one hand, and for siding, millwork, and cabinets on the other.

Weight

Weight, in addition to being important in itself, is generally a reliable index of strength. A heavy piece of wood is generally stronger than a lighter piece of the same moisture content and size, whether it is of the same or of a different species.

Wood weights, as commonly expressed, are either in the green or in the air-dry condition. Green weight of wood is the weight before any drying takes place; air-dry weight of wood refers to the weight after drying by exposure to atmospheric conditions for a time, either outdoors or in unheated sheds. The classification in table 1, column 3, is based on the air-dry condition.

Freedom From Shrinkage and Swelling

Most materials change in dimension with changes in temperature or moisture. Wood, like many other fibrous materials, shrinks as it dries and swells as it absorbs moisture. As a rule, however, much shrinking and swelling of wood in structures can be avoided by using wood that has been dried to a suitable moisture content.

For most species, the shrinkage or swelling in width of a flat-grained or plainsawed board[3] is often approximately twice that of an edge-grained or quartersawed board[5] of the same width (see fig.1-1). Edge-grained boards or other items cut from a species with high shrinkage characteristics will therefore prove as satisfactory as flat-grained boards or items cut from species with lower shrinkage characteristics. The normal wood of all species shrinks or swells very slightly along the grain (lengthwise).

The classification according to the amount of shrinkage (table 1, column 4) generally compares the performance of woods of various species. It does not tell the user the whole story of the shrinking and swelling of different species in service. Shrinkage of wood begins when moisture in the wood is removed by drying below the fiber saturation point (approximately 30 percent moisture content). When wood reaches a moisture content of 15 percent, about one-half of the total shrinkage has occurred. The moisture content of wood in service constantly changes since it adjusts to corresponding changes in surrounding atmospheric conditions.

The moisture content of woodwork installed within heated buildings reaches a low point during the heating season and a high point during the summer. The moisture content at the time of installation should be near the midpoint of

---

[3] Lumber that is cut tangent (roughly parallel) to the annual growth rings of the tree produces plainsawed boards in hardwoods and flat-grained or slash-grained boards in softwoods. Lumber that is cut at right angles to the annual rings, or parallel to the radius of the log, produces quartersawed boards in hardwoods, and edge-grained or vertical-grained boards in softwoods.

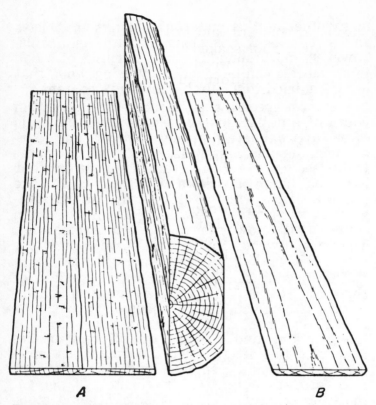

FIGURE 1-1. Grain depends on how lumber is cut from log. Board A is quartersawed or edge-grained. Board B is plainsawed or flat-grained.

this range. If this rule is followed, slight shrinkage will occur during some seasons and a slight swelling during others.

In a large portion of the United States, moisture content of interior woodwork in use varies between 5 and 10 percent with an average of about 8 percent. Siding and exterior trim in this same area varies between 9 and 14 percent, averaging about 12 percent. The dry southwest areas average below these percentages and the damp southern coastal States average above.

Plywood is relatively free from shrinkage and swelling as compared to solid wood because its construction generally consists of alternate laminations of veneers laid with grain at 90° to each other. From soaked to ovendry condition, the shrinkage of plywood in length and width is generally quite uniform and ranges from only about 0.2 to about 1.2 percent. After manufacture, plywood has a low moisture content and normally does not require drying out before use.

Methods of determining whether wood is dry enough for use are discussed later.

Freedom From Warping

The warping of wood is closely allied with shrinkage. Lumber that is crossgrained, or is from near the pith (core) of the tree, tends to warp when it shrinks. Classification of species according to their tendencies to warp and twist during seasoning, and as a result of changes in atmospheric conditions once the wood is dry, is listed in table 1, column 5. Warping can be reduced to a minimum by the use of edge-grained dry material.

The combined characteristics of warping and shrinkage determine the ability of wood to remain flat, straight, and not change size while in use. These qualities are desired in practically all uses. They are especially important in furniture, cabinetwork, window sash and frames, doors, and siding. Proper seasoning is important, but good construction details outlined later for preventing shrinkage also effectively prevent warping.

Ease of Working

Wood is generally easy to cut, shape, and fasten with ordinary tools directly on the building site. For some purposes the difference between woods in ease of working is negligible, but for others it may decidedly affect the quality and cost of the finished job. In general, ease of working is of first importance to the worker and indirectly to the one who pays the bill. Fabrication and assembly at the factory of cabinets,

TABLE 1-1. Broad classification of woods according to characteristics and properties A, woods relatively high in the particular respect listed; B, woods intermediate in that respect; C, woods relatively low in that respect. (Letters do not refer to lumber grades.)

Kind of wood	Working and behavior characteristics												Strength properties			
1	Hardness	Weight, dry	Freedom from shrinkage and swelling	Freedom from warping	Ease of working	Paint holding	Nail holding	Decay resistance of heartwood	Proportion of heartwood	Amount of figure	Freedom from odor and taste (dry)	Bending strength	Stiffness	Strength as a post	Toughness
	2	3	4	5	6	7	8	9	10	11	12	13	14	15	16
Ash: Black	B	B	C	B	C	C	A	C	C	A	A	B	B	C	A
White	A	A	B	B	C	C	A	C	C	A	A	A	A	A	A
Aspen	C	C	B	B	A	A	C	C	B	C	A	C	B	C	C
Basswood	C	C	C	B	A	A	C	C	C	C	A	C	B	C	C
Beech	A	A	C	C	C	B	A	C	B	B	A	A	A	B	A
Birch	B	B	C	B	C	B	A	C	C	B	A	A	A	B	A
Cedar: Eastern red	B	B	A	A	B	A	B	A	B	B	C	B	C	B	B
Northern white	C	C	A	A	A	A	C	A	C	C	B	C	C	C	C
Southern white	C	C	A	A	A	A	C	A	A	C	B	C	C	C	C
Western red	C	C	A	A	A	A	C	A	A	B	C	C	C	B	C
Cherry	B	B	B	A	B	B	A	C	B	B	B	A	A	A	B
Cottonwood	C	C	A	C	B	A	C	C	C	A	B	C	B	C	C
Cypress	B	B	A	B	B	A	B	A	B	A	B	B	B	B	B
Douglas-fir	B	B	B	B	B	B	A	B	A	A	C	A	A	A	A
Elm: Rock	A	A	B	B	C	C	A	C	B	A	A	A	A	A	A
Soft	B	B	B	C	C	C	A	C	B	A	A	B	B	B	A

54

Species														
Fir: Balsam	C	C	B	B	B	C	C	B	B	B	A	C	C	C
White	C	C	C	A	B	C	C	C	C	C	A	A	B	B
Gum	B	B	C	C	C	C	B	A	C	B	B	B	A	B
Hackberry	B	B	C	B	C	B	C	C	C	B	A	B	C	A
Hemlock: Eastern	A	C	B	B	B	B	B	C	B	B	A	B	B	C
West coast	B	A	A	B	B	B	A	C	B	B	B	A	A	B
Hickory	A	A	C	A	B	B	B	C	A	A	A	A	A	A
Larch: Western	B	A	B	B	C	A	C	A	A	B	O	A	A	B
Locust	A	A	B	B	B	C	A	A	A	A	B	A	A	A
Magnolia	B	B	B	B	B	B	C	A	A	A	A	A	A	B
Maple: Hard	A	A	B	B	B	C	A	C	A	A	B	A	A	A
Soft	B	B	B	B	B	C	B	C	A	A	A	A	A	A
Oak: Red	A	A	B	B	B	C	C	B	A	A	B	A	A	A
White	A	A	B	B	B	C	C	B	A	A	B	A	A	A
Pecan	A	A	B	B	C	A	A	C	A	A	B	A	A	A
Pine: Idaho white (western)	C	C	B	A	C	A	A	C	C	C	C	B	B	C
Lodgepole	C	C	B	C	C	C	B	B	A	O	C	B	B	B
Northern white (eastern)	C	C	A	A	A	A	C	C	C	O	C	B	C	B
Ponderosa	C	C	B	B	B	B	B	A	A	C	A	B	B	B
Southern yellow	B	A	C	B	A	B	C	C	B	C	A	B	B	C
Sugar	C	C	B	B	A	A	A	B	B	C	A	B	B	B
Poplar	C	C	B	B	B	B	A	B	B	B	A	B	C	C
Redwood	B	C	B	B	B	B	A	C	C	B	A	B	B	A
Spruce: Eastern	C	C	B	B	B	B	O	O	O	C	A	A	A	A
Engelmann	C	C	A	A	B	A	C	O	B	O	A	A	B	B
Sitka	O	C	B	A	C	B	B	B	C	B	A	B	B	C
Sycamore	B	B	B	B	C	B	B	A	C	B	A	B	B	B
Tupelo	B	B	B	B	C	B	B	C	O	O	A	B	B	C
Walnut	B	A	B	A	B	C	A	A	B	B	A	A	A	A

[a] Indicates general paintability and performance characteristics of edge-grained surfaces exposed to the weather.

55

TABLE 1-1. Con't

Kind of wood	Surface characteristics of common grades					Distinctive and principal uses
	Knots		Pitch defects	Other defects		
	Freedom from	Acceptance as to size	Freedom from	Freedom from	Acceptance as to size	
1	17	18	19	20	21	22
Ash: Black	A	B	A	B	B	Implements, cooperage, containers
White	A	B	A	B	B	Implements, containers, furniture
Aspen	A	A	A	B	B	Boxes, lumber, pulp, excelsior
Basswood	A	B	A	A	A	Woodenware, boxes, veneer
Beech	B	B	A	C	C	Flooring, furniture, woodenware
Birch	A	B	A	B	B	Flooring, furniture, millwork
Cedar: Eastern red	C	A	A	A	A	Posts, paneling, wardrobes, chests
Northern white	B	B	A	B	B	Poles, posts, tanks, woodenware
Southern white	A	B	A	A	A	Posts, poles, boat and tank stock
Western red	A	C	A	B	B	Shingles, siding, poles, millwork
Cherry	A	B	A	B	B	Furniture, woodenware, paneling
Cottonwood	A	A	A	A	A	Pulpwood, excelsior, containers
Cypress	A	B	A	B	B	Millwork, siding, tanks
Douglas-fir	B	B	B	B	B	Construction, plywood, millwork
Elm: Rock	B	B	A	B	B	Furniture, containers, veneer
Soft	B	B	A	C	C	Containers, furniture, veneer

Species						Uses
Fir: Balsam	C	A	—	B	B	Light construction, pulpwood
White	B	B	A	—	C	Light construction, containers
Gum	A	B	A	B	A	Millwork, containers, furniture
Hackberry	A	C	A	B	B	Furniture, veneer, containers
Hemlock: Eastern	B	B	A	C	C	Construction, containers
West coast	B	B	A	B	B	Construction, pulpwood, containers
Hickory	B	C	A	B	B	Handles, athletic goods, implements
Larch: Western	C	A	A	C	A	Construction, poles, ties, millwork
Locust	B	B	A	B	B	Poles, posts, insulator pins, ties, fuel
Magnolia	A	B	A	B	B	Furniture, veneer, containers
Maple: Hard	B	B	A	A	B	Flooring, furniture, veneer
Soft	A	C	A	A	A	Furniture, woodenware, fuel
Oak: Red	A	C	A	B	B	Flooring, furniture, veneer, posts
White	A	C	A	B	B	Furniture, cooperage, millwork, veneer
Pecan	A	C	A	C	C	Implement handles, flooring, pallets
Pine: Idaho white (western)	C	A	A	C	A	Millwork, construction, siding, paneling
Lodgepole	C	B	A	B	B	Poles, lumber, ties, mine timbers
Northern white (eastern)	C	A	A	B	B	Millwork, furniture, containers
Ponderosa	B	B	B	B	B	Millwork, construction, poles, veneer
Southern yellow	A	C	C	B	A	Construction, poles, siding, cooperage
Sugar	C	B	A	B	B	Millwork, patterns, construction
Poplar	A	B	A	A	A	Furniture, plywood, containers
Redwood	A	C	A	A	B	Siding, tanks, millwork
Spruce: Eastern	C	A	A	B	B	Construction, pulpwood
Engelmann	C	A	A	B	B	Light construction, poles, pulpwood
Sitka	B	B	A	B	B	Construction, millwork, containers
Sycamore	A	B	A	B	B	Furniture, veneer, cooperage
Tupelo	A	A	A	A	A	Containers, furniture, veneer
Walnut	A	B	A	A	A	Furniture, gunstocks, interior finish

windows, frames, doors, and other units have greatly reduced the time required for the skilled worker at the building site.

Harder and denser woods with high load-carrying capacity and wear resistance should not be passed over just because softer woods are easier to work; rather, a reasonable balance must be drawn in selecting wood for a specific use.

A skilled carpenter working with lumber that is well seasoned and manufactured can get good results from even the more difficult-to-work woods. An unskilled worker is more likely to get good results only from the softer woods. However, with portable power tools, jigs for installation of hinges and door locks, and other modern labor-saving methods, skill is no longer the major factor it was when hand tools were the only means of cutting and fitting on the job.

The classification of the more common woods according to their working qualities (table 1, column 6) is based on a combination of the hardness, texture, and character of the surfaces obtainable. Woods in the A class have soft, uniform textures and finish to smooth surfaces; woods in the C class are hard or nonuniform in texture and more difficult to surface without chipping the grain, fuzzing, or grain raising. The B class is intermediate.

Paint Holding

Good paint performance or ability of a wood surface to hold paint depends on three factors: (1) the kind of paint, (2) surface conditions and application factors, and (3) the kind of wood. The first two factors are discussed later under the section "How and When to Paint" only the kind of wood is discussed at this point.

Different woods vary considerably in painting characteristics, particularly for outdoor exposure. The ratings of the species in table 1, column 7, indicate generally their abilities to hold paint under exposure to the weather. The best species for exterior painting are the woods in A class, including such common ones as the cedars, redwood, ponderosa pine, or white pine.

Paint is more durable on edge-grained surfaces than on flat-grained surfaces. The edge-grained boards in B class woods usually have a better surface for painting than the flat-grained surfaces of A class woods.

Knots, particularly resinous ones, do not hold paint well and contribute to abnormally early paint failure. High content of pitch and resin will also detract from the paintability of wood unless the pitch is set adequately by proper high-temperature seasoning of the wood.

Class B and class C woods and plywood are best finished with pigmented stains that penetrate the wood surface and do not form a continuous film on the surface. Such stain finishes do not fail by cracking and peeling of the coating from the wood as does paint. The stains are also recommended for use on shingle and shake sidewalls and rough-sawn lumber and siding.

Nail Holding

As a rule, fastenings are the weakest link in all forms of construction and in all materials; therefore the resistance offered by the wood to the withddrawal of nails is important. Usually, the denser and harder the wood, the greater is the inherent nail-holding ability, assuming the wood does not split. The grouping of the commercial woods (table 1, column 8) according to their inherent nail-holding ability is based on tests that measured the force required to pull nails from wood.

The size, type, and number of nails have a marked effect on the strength of a joint. Fig. 2 illustrates good and poor nailing practices at the foundation wall. Correct placement of the

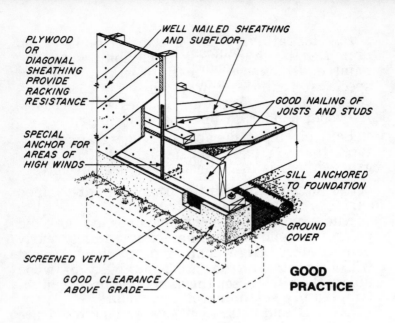
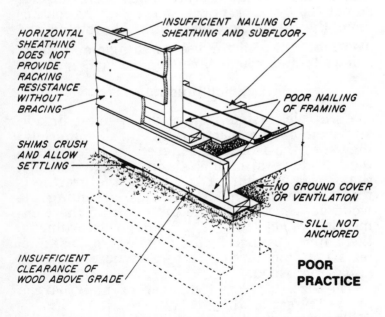

FIGURE 1-2. Good and poor construction practices at foundation.

nails is as important as the size and number. If high winds are general during severe storms, special fastenings should be used to resist these pressures.

The resistance of nails to withdrawal increases almost directly with their diameter; if the diameter of the nail is doubled, the holding strength is doubled, providing the nail does not split the wood when it is driven. The lateral resistance of nails increases as the $1\frac{1}{2}$ power of the diameter.

The nail most generally used in wood-frame construction is the common nail. However, galvanized and aluminum nails are used extensively in applying siding and exterior trim because these nails resist rusting. The galvanized nail is slightly better than the common bright nail in retaining its withdrawal resistance.

Superior withdrawal resistance has been shown by the deformed-shank nail, which is produced in two general forms, the annular-groove and the spiral-groove shanks. The annular-groove nail is outstanding in its resistance to static-withdrawal loads but not as good as the spiral-groove nail when subjected to racking loads. The spiral-groove nail is superior to the plain-shank nail in its resistance to withdrawal loads and is commonly used in construction of pole-type buildings.

Interior carpentry uses the small-headed finish nail, which can be set and puttied over.

The moisture content of the wood at the time of nailing is extremely important for good nail holding. If plain-shank nails are driven into wet wood, they will lose about three-fourths of their full holding ability when the wood becomes dry. This loss of holding power is so great that siding, barn boards, or fence pickets are likely to become loose when plain-shank nails are driven into green wood that subsequently dries. Thus the most important rule in obtaining good joints and high nail-holding ability is to use well-seasoned wood.

Prevention of Splitting.—The splitting of wood by nails greatly reduces their holding ability. Even if the wood is split only slightly around the nail, considerable holding strength is lost. Because of hardness and texture characteristics, some woods split more in nailing than do others. The heavy, dense woods, such as maple, oak, and hickory, split more in nailing than do the light-weight woods such as basswood, spruce, and balsam and white fir.

Predrilling is good practice in dense woods, especially when large diameter nails are used. The drilled hole should be about 75 percent of the nail diameter.

Woods without a uniform texture, like southern yellow pine and Douglas-fir, split more than do such uniform-textured woods as northern and Idaho white pine, sugar pine, or ponderosa pine.

In addition to predrilling, the most common means taken to reduce splitting is the use of small diameter nails. The number of small nails must be increased to maintain the same gross holding strength as with larger nails. Slightly blunt-pointed nails have less tendency to split wood than do sharp-pointed nails. Too much blunting, however, results in a loss of holding ability.

Decay Resistance

Every material has its distinctive way of deteriorating under adverse conditions. With wood it is decay. Wood will never decay if kept continuously dry or continuously under water. Fortunately, most wood in ordinary buildings is in dry situations and therefore not in danger of decay. It is only in certain parts of the buildings that decay resistance is important, such as areas where wood may become damp or where it touches or is embedded in the ground.

To protect wood from decay, there are three things which can be done, either singly or in combination: (1) make sure it is dry when installed and kept dry in service; (2) use the heartwood of a decay-resistant species where occasional wetting and drying can be expected; or (3) use wood that has been given a good preservative treatment for places where moisture is certain to get in, as from contact with the soil or because of poor drainage or ventilation.

The different kinds of wood are classified in accordance with their natural decay resistance in table 1, column 9. This classification applies solely to the heartwood, because sapwood of all species in the untreated condition has low decay resistance. Also, this classification deals only with averages, and exceptions frequently occur because of variations in the wood itself and because of differences in the kinds of fungi that cause the decay.

Further information on avoiding decay, including the use of preservative-treated material, is contained in the later section on "Preventing Decay."

Proportion of Heartwood

When selecting untreated wood for use where the decay hazard is high, one must consider the heartwood content, because only the heartwood is decay resistant. When the sapwood of the species of tree is normally narrow, as it is in the woods rated as class A in table 1, column 10, the lumber runs high in heartwood content even without special selection. When the sapwood is normally wide, as in woods rated as class C and even in class B in column 10, the commercial run of lumber contains considerable sapwood.

To obtain decay-resistant lumber, even in the species classed as A in decay resistance in column 9, it is necessary to eliminate the sapwood by special selection. Specially selected building lumber, sold in "all-heart" grades, is procurable in cypress, redwood, western red cedar, and

Douglas-fir. However, all-heart grades in southern yellow pine are special and are not easily obtainable.

Figure

Figure is due to various causes in different woods. In woods like southern yellow pine and Douglas-fir, it results from the contrast between springwood and summerwood in growth rings; in oak, beech, or sycamore, it results from the flakes or rays in addition to the growth rings; in maple, walnut, and birch it results from wavy or curly grain; and in gum it results from infiltrated coloring matter.

Except where the figure in wood results from flakes or rays, it is more pronounced in flat-grained lumber than in edge-grained. Figure resulting from wavy or curly grain or from infiltrated color does not occur in all lumber of a given species, but only in lumber from occasional logs. To be certain of getting figured lumber in maple, walnut, or gum, special selection is necessary.

Woods with outstanding knots, such as pine and cedar, or with other unique characteristics such as those of pecky cypress, or "white speck" Douglas-fir, are often selected because of their novel patterns. The finish selected for these types of wood tends to accentuate rather than obscure the knots or other features. The advantage of figure or color may appear in the interior trim, in the floor, or in a wood-paneled wall.

The color of the wood has a decided influence on the figure. However, stains are so commonly and easily applied to most woods that natural color is usually not the first consideration, except where a very light color is desired.

A broad classification of the important kinds of lumber, from the standpoint of the amount of figure they contain, is shown in table 1, column

11. Woods classed as A are highly figured, and an ordinary commercial run will have a pronounced figure. Class B woods have more modulated figures and sometimes require special selection to obtain the desired figure. Class C woods are seldom satisfactory where figure is desired.

Freedom From Odor and Taste When Dry

None of the common woods has sufficient odor to prevent its satisfactory use in building construction. It is only when the wood is used for food containers that odor and taste are critical. When green, all woods have some odor and will impart a woody taste to very susceptible foods. After the woods are dried, however, many have practically no odor or taste. The principal objection to odor and taste in wood is that they contaminate food they touch, especially butter and cheese. On the desirable side, the aromatic odor of the cedars is prized for such uses as clothes closets and chests.

The woods grouped in class A in table 1, column 12, are suitable for use in contact with foods that absorb odors. The woods in class C have a strong resinous or aromatic odor and are unsuited for use in direct contact with foods that absorb odors. Woods in class B cannot be used in contact with very susceptible foods, like butter, but they do not have the strong odor and taste of the aromatic and resinous woods.

Bending Strength

Bending strength is a measure of the load-carrying capacity of members that are ordinarily used in a horizontal or moderate slope position and rest on two or more supports. Examples of members in which bending strength is important are rafters, ceiling and floor joists, beams or girders, purlins, bridge stringers, and scaffold platforms.

Even though a species is low in bending strength, it may still be selected for uses where this property is essential. However, larger sizes are then required to obtain the same load-carrying capacity.

A small increase in the depth of a beam produces a much greater percentage increase in bending strength than it does in volume. An increase of 1 inch in the depth of a 10-inch beam (from 10 to 11 inches) will increase its volume 10 percent, whereas the bending strength of the beam is increased 21 percent. An increase in the width of a beam, however, increases the bending strength by the same percentage as the volume. An increase of 1 inch in a beam 10 inches deep will increase both bending strength and volume by 10 percent.

No simple rule can be given to determine the size of girder, joist, or plank required to carry a given load. However, lumber associations have prepared books that contain tables of safe loads for given spans, sizes, species, and spacings.

The softwoods in class A in table 1, column 13, such as Douglas-fir, southern yellow pine, and western larch, dominate the structural field. They are used both for heavy construction (barns and bridges) and light construction (dwellings and small farm structures). In heavy construction, softwoods in class B are used only occasionally. In light construction, softwoods in class B such as white fir, hemlock, and Idaho white pine are used extensively. Their light weight and ease of working enable them to compete with the stronger woods.

Woods in class C are relatively unimportant in the structural field as they are seldom used in heavy construction and only occasionally in light construction. The hardwoods in classes A and B have largely dropped out of the construction field, not because they are unsuited to the use, but because of their value for uses with more exact requirements—furniture, flooring, and veneers in plywood.

Stiffness

Stiffness is a measure of the resistance to bending or deflection under a load. In the floor and ceiling joists of a house, stiffness is more important than actual breaking strength,[5] because it is deflection or sag that must be reduced to a minimum to avoid plaster cracks in ceilings and vibration in floors. Stiffness is important also in shelving, ladder rails, beams, and long, slender columns.

Whereas stiffness is of great importance in floor joists, the advantages of using a relatively stiff species will be lost if the members are not fully dry at time of installation, so the fastenings and bracing hold well. Straight, well-seasoned joists of a species that is relatively low in stiffness may give better results than an inherently stiff wood that is green or carelessly installed. If the wood is sufficiently dry and the installation is good, however, species differences with respect to stiffness are important.

Differences in stiffness between species may be compensated for by changing the size of members. Depth and length of members have a greater effect on their stiffness than on other strength properties. For example, a change of 1/32 inch in the thickness of a 25/32-inch board produces a change of 12 percent in the stiffness of the board laid flat in a floor. A 10-inch joist has about one-fourth more wood in it than an 8-inch joist, but set on edge in a building it is more than twice as stiff.

The species are classified by stiffness in table 1, column 14. Softwoods in class A and class B dominate the uses where stiffness is the most important requirement. When woods in class C are used where stiffness is desired, it is because

---

[5] Breaking strength refers to the load required to break a material, while stiffness refers to its ability to sustain loads with a minimum of deflection or sag.

other properties are more important. The woods in class A have the highest stiffness, but they are heavier and harder than those in Class B.

Light weight is quite commonly desired in combination with stiffness. The softwoods meet this requirement much better than the hardwoods, and softwoods in class B are often chosen in preference to those in class A because the weight of the latter excludes them.

Strength as a Post

Posts or "compression members" are generally square or circular in cross section, usually upright, and support loads that act in the direction of the length. Strength in compression is an essential requirement for posts supporting beams in a basement or crawl space, for supports of root cellars, for storage bins, and for posts in similar heavy construction where the length is less than 11 times the smallest dimension. It is not important in fenceposts, which carry no loads.

In small buildings the size requirements of posts, with the smallest dimension less than one-eleventh of the length, are determined by bearing area, stiffness, and stability, rather than by actual compressive strength. Therefore it is necessary to use posts large enough to carry much greater compressive loads than are ever placed upon them. No great consideration need therefore be given to compressive strength endwise in selecting wood for small houses.

Where exceptionally heavy loads are involved as in supports for bins or underground cellars, consideration should be given to the compressive strength of different woods as shown in table 1, column 15. Even where compressive strength is an important requirement, the woods in any class may be safely used, provided the lower strength of class B and C woods is compensated for by using timber of larger cross-sectional area.

When the length of the post or column is greater than 11 times the smallest dimension, stiffness becomes an important factor in determining the load-carrying ability. Unbraced supports, such as squared posts in machine sheds or barns and poles in pole-type structures, are generally so slender that they should be judged by their stiffness rather than their compressive strength.

Toughness

Toughness is a measure of the capacity to withstand suddenly applied loads. Hence, woods high in shock resistance are adapted to withstand repeated shocks, jars, jolts, and blows, such as are given ax handles and other tool handles. The heavier hardwoods—hickory, birch, oak, maple, and ash—are so much higher in shock resistance than the toughest of the softwoods that these hardwoods are used almost exclusively where an exceptionally tough wood is required.

None of the softwoods in table 1, column 16, is grouped in class A in toughness, and few hardwoods (aspen, cottonwood, and basswood) fall in class C. The woods in class A completely dominate the uses where toughness is the outstanding requirement, and hickory dominates class A.

Toughness is a desirable property in uses other than those in which it is required. Tough woods give more warning of failure than do brash woods. It is, therefore, a factor in beams and girders where heavy loads are applied. The selection of class C woods should normally be avoided for these two uses.

Surface Characteristics of Common Grades of Lumber

Lumber is purchased by home owners because of its appearance as well as its working characteristics and strength properties. The appearance is dependent largely on the grade, and there is some degree of uniformity in the appearance of the same grade in different woods. Different woods are more uniform in appearance in the Select grades than in Common or Dimension grades because most knots, pitch pockets, and the like are eliminated from the Select grades.

In the Common grades, however, where knots and similar surface features are allowable, there are differences in the same grade of different woods. These differences affect the appearance of the wood and at times its suitability for a use. For example, the number of knots and like features in a board averages, in different species, from about 5 to 20 per 8 board feet regardless of grade. Second and third grade Common Boards are selected for greatest utility. Fourth and lower grades permit moderate utility. Grades for the various species for Board and Dimension lumber are outlined in a following section.

Table 1, therefore, includes in columns 17 to 21 a classification of various woods according to the size and number of the more important surface features found in the Common grades. The woods classed as A have the least number of these surface features and are generally the most desirable. Those classed as C have the most knots and other surface features and are generally not as desirable, unless these features are used for an architectural or decorative effect such as knotty paneling. (These letters do not correspond to lumber grades.)

Distinctive and Principal Uses

The distinctive and principal uses to which a wood is put are indicative of its properties. Distinctive and principal uses are those to which a wood is most generally fitted. The fact that a wood's distinctive use is for gunstocks, for ax handles, for woodenware, or for fenceposts tells one who is familiar with these uses much more about the wood than does a verbal description or a table of properties, unless he has been trained to combine and evaluate the properties. A knowledge of the requirements for ax handles obtained from actual experience gives a good idea of the combination of toughness, breaking strength, stiffness, and texture to be found in a wood used for that purpose.

The distinctive and principal uses listed in table 1, column 22, therefore supplement the data on properties and aid in visualizing the general character of the wood.

LUMBER GRADES

Lumber grading rules are formulated and published by associations of lumber manufacturers or by official grading and inspection bureaus.

Softwood Lumber

Finish or Select grades.—Finish or Select grades of lumber generally are named by the letters A, B, C, and D. The A and B grades are nearly always combined as B and Better, so that only three grades are in practical use.

Thus, in lumber for interior and exterior finishing or other similar uses, only B and Better (first grade), C (second grade), and D (third grade) in softwoods need to be considered. However, considerable knotty pine and cedar in third grade are selected for use as paneling.

Common boards.—Grade names for common boards are not uniform for all softwood species. For example, in redwood boards Select is the first grade, Construction the second, Merchantable the third, and Economy the fourth. With such woods as Douglas-fir, west coast hemlock,

Sitka spruce, and western redcedar, the grade designations are Select Merchantable, Construction, Standard, Utility, and Economy. A different set of board grades described by the Western Wood Products Association bears the names 1 Common, 2 Common, 3 Common, 4 Common, and 5 Common. The same set of grade names and descriptions is used in the Northeast and Lake States for species such as eastern spruce, balsam fir, red and jack pine, eastern hemlock, and northern white cedar. For the southern pines, the board designations are No. 1, No. 2, No. 3, and No. 4.

Dimension lumber.—Light Framing (2 to 4 inches thick, 2 to 4 inches wide) and Joists and Planks (2 to 4 inches thick, 6 inches and wider) are graded for strength to a common set of grade names and descriptions under all six softwood grading rules published in the United States, and under the National Lumber Grades Authority in Canada. The light framing grades are Construction, Standard, and Utility, and the Joist and Plank grades are Select Structural, No. 1, No. 2, and No. 3. There is also a Structural Light Framing category for roof truss and similar applications that has the same grade names as for Joists and Planks. Load-carrying design values vary by species and use category; therefore, it is important to note that common grade names do not imply equal strength or stiffness.

Trade practices.—It has been the practice for the lumber retailer to quote prices and make deliveries on the basis of local grade classification or on his own judgment of what the user needs or will accept. However, there is a growing practice to put indelible marks on all building lumber at the sawmill, stating the grade, species, size, degree of seasoning, and identity of the supplier. The Federal Housing Administration (FHA) and most code authorities require that framing lumber used in the construction of FHA-insured units be so grade-marked.

The softwoods are graded to meet fairly definite building requirements. Select grades of softwoods are based on suitability for natural and paint finishes: A Select and B Select or B and Better are primarily for natural finishes, and C Select and D Select are for paint finishes. The Utility Board and Dimension grades are based primarily on their suitability for general construction and general utility purposes as influenced by the size, tightness, and soundness of knots.

Hardwood Lumber

The wood of the hardwood trees is graded on the basis of factory grades more than for building requirements. Factory grades take into account the yield and size of cuttings with one clear face that can be sawed from the lumber. The two highest factory grades are known as Firsts and Seconds, and are usually sold combined.

Hardwoods for construction are grouped into three general classes: Finish, Construction and Utility Boards, and Dimension. A Finish has one face practically clear, while B Finish allows small surface checks, mineral streaks, and other minor variations.

Construction Boards and Utility Boards have No. 1, No. 2, and No. 3 grades and are based on the amount of wane, checks, knots, and other defects present in each board.

Dimension grades (2 inches thick) are classed as No. 1 and No. 2 depending on the number of defects.

Strength Factor

The ordinary grades of building or so-called yard lumber are based on the size, number, and location of the knots, slope of grain, and the like more than on the strength of the clear wood.

Common softwood boards used in conventionally constructed houses and other light-frame structures are not related directly to the strength of the unit itself. Rather, sheathing, subflooring, and roof boards supplement the framing system and may also add to the rigidity of the structure.

The main purpose of boards used in the construction of a building is as a covering material. They also facilitate nailing for siding, flooring, and roofing materals. For these purposes they must have some nail-holding properties as well as moderate strength in bending to carry loads between the frame members. Ordinarily, third and fourth grade boards are adequate for this purpose.

Finish and Select grade softwood boards are selected for their appearance rather than strength. They are used mainly for trim and finish purposes, and consequently the grade is chosen based on the type of finish used—natural, stained, or painted.

Softwood Dimension lumber is selected because of its strength and its stiffness. Therefore, the size, number, and location of knots are important and related directly to the intended use. In house floors and walls, for example, where construction is designed to minimize vibration and deflection so far as possible, stiffness rather than breaking strength is most important. Generally, grade affects strength more than stiffness; the lower the grade, the lower the strength.

Finishing and Appearnace Factor

The finishing and appearance of wood is normally associated with the various Board grades rather than Dimension grades. With varnish and natural finishes, A and B Select in softwoods (commonly sold as B and Better) and A Finish in hardwoods assure the best appearance. Some pieces in the B and Better grade are practically clear, although the average board contains one or two small surface features that preclude calling it Clear.

Where the very smoothest appearance is not required, second Finish grade in softwoods and hardwoods gives good satisfaction. The number of knots, pitch pockets, and other nonclear features per board in C Select averages about twice that of B and Better; the proportion of these features that are small knots is greater in C Select than in B and Better. Because of its decorative effects, knotty lumber selected from the first and second Common board grades is frequently in demand for paneling.

For painting where wood is not exposed to the weather, the surface features permitted in the second Finish grade are such that they can be well covered by paint if the priming is properly done. The third Finish grade, with some cutting out of defects, gives almost as good quality as the second grade. Some of the natural features and manufacturing imperfections are not much more numerous in the third grade than in the second grade, but the number and size of the knots are considerably greater, and often the back of the pieces is of lower quality. Where smoothest appearance at close inspection is required under exposure to the weather, first Finish grade gives the best results.

For painted surfaces that do not receive close inspection (barns, summer cottages, and the like) and where protection against the weather is as important as appearance, the first and second Board grades are satisfactory. The larger knots and pitch pockets in the second grade Common softwood boards do not give as smooth and lasting a painted surface as do the smaller ones in the first grade, but the general utility is good.

Tightness Factor

First grade Common softwood and Utility hardwood boards are suitable for protection against rain or other free water beating or seeping through walls or similar construction. These and the Finish grades are usually kept drier at the lumber yards than are the lower grades, and will therefore shrink and open less at the joints if used without further drying. Where only tightness against leakage of small grain is required in a granary or grain bin, second grade boards may be used with a small amount of cutting to eliminate knotholes. When used as sheathing with good building paper, second grade boards are satisfactory even though knotholes and other similar openings do occur.

Wear-Resistance Factor

Edge-grained material wears better than flat grain, narrow-ringed wears better than wide-ringed, and clear wood wears more evenly than wood containing knots. The first Finish grades in softwoods and hardwoods ordinarily contain very few defects and withstand wear excellently. The second grade in softwoods and in hardwoods sufficiently limits knots and surface characteristics to assure good wearing qualities. Third Finish grade and first grade boards limit the size and character of knots, although not the number, and are satisfactory where maximum uniformity of wear is not required.

Decay-Resistance Factor

Any natural resistance to decay that a wood may have is in the heartwood. The decay resistance of the species so far as affected by grade therefore depends upon the proportion of heartwood in the grade. While this is true of all species, it is of practical importance only in woods with medium or highly decay-resistant heartwood.

The lower grades usually contain more heartwood than do the Select grades. If decay resistance is really needed for the purpose at hand, the first and second board grades are more decay-resistant than are the Finish grades, except in the case of the special Finish grades known as All Heart.

The full decay resistance of grades below the second grade is reduced by the presence of decay that may have existed in the tree or log before it was sawn into lumber. Under conditions conducive to decay, such original decay may spread, although some types of decay, notably peck in cypress, red heart in pine, and white pocket in Douglas-fir, are definitely known to cease functioning once the lumber is properly seasoned.

Price Factor

The spread in price between Select Finish and Utility Board grades varies considerably from time to time, depending upon supply and demand. The cost of the lower Select grades is substantially greater than the upper Board grades of softwoods. With such a difference in price it is obviously important not to buy a better grade than is needed. *Any tendency to buy the best the market offers for all uses is wasteful of both lumber and money,* for in uses such as sheathing, the lower and cheaper grades will render as long and satisfactory service as the higher priced grades.

The price spread between the combined grade of first and second Finish grades and the Common grades of hardwoods is also large. This is of minor importance to builders because most of the hardwood purchased by them has already been manufactured into some form of finished product, such as flooring or interior trim.

Roughly the combined grade of first and second Finish may have a market value from 50 to 100 percent greater than that of the highest Common grade, and contain from 25 to 50 percent more clear-face cuttings of the sizes specified in the grading rules. If large clear-face pieces are required, they can best and possibly only be obtained from the first and second grades. But if only medium-sized or small clearface pieces are required, they can be obtained from the Common grades.

STANDARD LUMBER ITEMS

Lumber is sold as a number of standard general-purpose items and also as certain special-purpose items. Retail lumberyards carry all the general-purpose items and the more important of the special-purpose items. Some lumber items can be obtained only in the upper grades, and others only in the lower. Few items are made in a complete range of grades. A brief description of framing and dimension, boards and sheathing, flooring and siding, and other lumber and related items commonly carried by most retail lumberyards is given later in this section.

Many lumberyards carry stock items in wood species besides those common to the United States. Larger lumber companies may also have their own sash and door plants and can make to order any wood unit listed in the plans or specifications of frame buildings. The popularity of the wood truss has also brought about the fabrication of these items at many lumberyards.

Dressed Thicknesses and Widths of Lumber

Lumber as ordinarily stocked in retail yards is surfaced (dressed) on two sides and two edges. This is to make the lumber ready to use and uniform in size without further reworking, and also to avoid paying transportation costs on material that would have to be cut off on the job. The amount that is reasonable and desirable to dress off has varied considerably in the past and has been the subject of some controversy and misunderstanding among producing and consuming groups. American lumber standards have been set up by the lumber trade with the assistance of Government agencies in such a way as to largely take care of the situation.

American lumber standards and common trade practices now provide dressed sizes as summarized in tables 2 and 3 which are taken from the American Softwood Lumber Standard Voluntary Product Standard PS 20-70. The column designated nominal size shows the dimensions according to which lumber is usually described; the last column shows the actual dimensions of lumber when it is sold surfaced.

When the dimensions of dressed lumber are less than those shown in the table for the actual sizes enumerated, the lumber is known as substandard. Items of some woods are commonly sold in substandard sizes. It is well to check the dimensions before selecting a wood so that allowance can be made in both price and utility for substandard sizes or proper credit given for oversizes.

Framing and Dimension

Dimension is primarily framing lumber, such as joists, rafters, and wall studs. It also comprises the planking used for heavy barn floors. Strength, stiffness, and uniformity of size are essential requirements. Framing or Dimension lumber is stocked in all lumberyards but often in only one or two of the general-purpose construction woods—Douglas-fir, southern yellow pine, white fir, hemlock, or spruce. It is usually a nominal 2 inches thick, dressed one or two sides to $1\frac{1}{2}$ inches dry (table 2). It is nominally 4, 6, 8, 10, or 12 inches in width, and 8 to 20 feet long in multiples of 2 feet. Dimension thicker than 2 (up to 5) inches and longer than 20 feet is manufactured only in comparatively small quantities.

Perhaps the one most suitable grade for permanent construction wall framing, based on economy and performance, is the third grade in the various species. The grade most generally

Table 1-2. Nominal and minimum-dressed sizes of boards, dimension, and timbers. (The thicknesses apply to all widths and all widths to all thicknesses.)

ITEM	THICKNESSES			FACE WIDTHS		
	NOMINAL	Minimum Dressed		NOMINAL	Minimum Dressed	
		Dry[6]	Green[6]		Dry[6]	Green[6]
		Inches	*Inches*		*Inches*	*Inches*
Boards[7]				2	1½	1 9/16
				3	2½	2 9/16
				4	3½	3 9/16
				5	4½	4⅝
	1	¾	25/32	6	5½	5⅝
				7	6½	6⅝
	1¼	1	1 1/32	8	7¼	7½
				9	8¼	8½
	1½	1¼	1 9/32	10	9¼	9½
				11	10¼	10½
				12	11¼	11½
				14	13¼	13½
				16	15¼	15½
Dimension				2	1½	1 9/16
				3	2½	2 9/16
				4	3½	3 9/16
	2	1½	1 9/16	5	4½	4⅝

Dimension	2½ 3 3½	2 2½ 3	2 1/16 2 9/16 3 1/16	6 8 10 12 14 16	5½ 7¼ 9¼ 11¼ 13¼ 15¼	5⅝ 7½ 9½ 11½ 13½ 15½
Dimension	4 4½	3½ 4	3 9/16 4 1/16	2 3 4 5 6 8 10 12 14 16	1½ 2½ 3½ 4½ 5½ 7¼ 9¼ 11¼	1 9/16 2 9/16 3 9/16 4⅝ 5⅝ 7½ 9½ 11½ 13½ 15½
Timbers	5 & Thicker		½ Off	5 & Wider		½ Off

[6] "Dry" lumber has been dried to 19 percent moisture content or less; "green" lumber has a moisture content of more than 19 percent.

[7] Boards less than the minimum thickness for 1 inch nominal but ⅝ inch or greater thickness dry (11/16 inch green) may be regarded as American Standard Lumber, but such boards shall be marked to show the size and condition of seasoning at the time of dressing. They shall also be distinguished from 1-inch boards on invoices and certificates.

TABLE 3. Nominal and minimum dressed dry sizes of finish, flooring, ceiling, partition, and stepping at 19 percent maximum moisture content

(The thicknesses apply to all widths and all widths to all thicknesses except as modified)

[8] For nominal thicknesses under 1 inch, the board measure count is based on the nominal surface dimensions (width by length). With the exception of nominal thicknesses under 1 inch, the nominal thicknesses and widths in this table are the same as the board measure or count sizes.

[9] In tongued-and-grooved flooring and in tongued-and-grooved and shiplapped ceiling of 5/16-inch, 7/16-inch, and 9/16-inch dressed thicknesses, the tongue or lap shall be 3/16 inch wide, with the over-all widths 3/16 inch wider than the face widths shown in the table above. In all other worked lumber of dressed thicknesses of 5/8 inch to 1¼ inches, the tongue shall be ¼ inch wide or wider in tongued-and-grooved lumber, and the lap of 3/8 inch wide or wider in shiplapped lumber, and the over-all widths shall be not less than the dressed face widths shown in the above table plus the width of the tongue or lap.

ITEM	THICKNESSES		FACE WIDTHS	
	NOMINAL[8]	Minimum Dressed	NOMINAL	Minimum [8] Dressed
		Inches		*Inches*
Finish	3/8	5/16	2	1½
	½	7/16	3	2½
	5/8	9/16	4	3½
	¾	5/8	5	4½
	1	¾	6	5½
	1¼	1	7	6½
	1½	1¼	8	7¼
	1¾	1 3/8	9	8¼
	2	1½	10	9¼
	2½	2	11	10¼
	3	2½	12	11¼
	3½	3	14	13¼
	4	3½	16	15¼
Flooring[9]	3/8	5/16	2	1 1/8
	½	7/16	3	2 1/8
	5/8	9/16	4	3 1/8
	¾	¾	5	4 1/8
	1¼	1	6	5 1/8
	1½	1¼		
Ceiling[9]	3/8	5/16	3	2 1/8
	½	7/16	4	3 1/8
	5/8	9/16	5	4 1/8
	¾	11/16	6	5 1/8
Partition[9]	1	23/32	3	2 1/8
			4	3 1/8
			5	4 1/8
			6	5 1/8
Stepping[9]	1¼	¾	8	7¼
	1½	1¼	10	9¼
	2	1½	12	11¼

suitable for joists and rafters for permanent and first-class construction is the second grade of the various species. Satisfactory construction is possible with lower grades, but pieces must be selected and there is considerably more cutting loss. Many species have structural grade classifications that may be used for trusses and other structural components. These structural grades allow greater loads than do equal spans of the lower grades.

Boards or Sheathing

Boards are a general-purpose item used most often to cover framing members as flooring, roofing, and wall sheathing. They are available at all lumberyards in one or more kinds of wood most frequently used in building construction. Boards are usually of nominal 1-inch thickness, dressed on two sides to $3/4$-inch dry thickness, and are usually manufactured in all grades from first to fifth (table 3). However, as sheathing material, the third and fourth grades are most often used.

Boards or sheathing are manufactured in a number of patterns. They may be square-edged (surfaced on four sides), generally supplied in nominal 4-, 6-, and 8-inch widths. They are also available in dressed and matched pattern (tongued-and-grooved) and in shiplap (fig. 1-3). Dressed and matched material is most commonly sold in 6-inch widths, and shiplap in 8-, 10-, and 12-inch widths. In addition to sheathing and subflooring, boards are used for rough siding, barn boards, and concrete forms. The advent of the pole-type construction has developed the need for center-matched sheathing in 2- by 6-inch nominal size. Many lumber companies stock this item preservative-treated.

Siding

Siding, as the name implies, is made and generally used for exterior coverage. It is produced

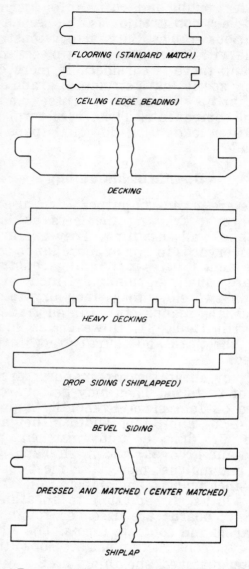

FIGURE 1-3. Typical patterns of lumber.

in several general patterns: bevel siding, drop siding, and V-edge siding or paneling (fig. 1-3). Bevel siding is ordinarily stocked in Clear, A grades and B grades of redwood, western red cedar, hemlock, white and ponderosa pine, and spruce. Drop siding is stocked in C and Better and No. 2 Common grades, and V-edge paneling in Clear, and C and Better. Additional species used for drop and V-edge siding are Douglas-fir and southern pine.

Other exterior lumber coverings include "boards and battens" and other combinations for vertical application. These are commonly rough-sawn boards in Finish grades that are given a stain finish.

Bevel siding (fig. 1-3) is commonly supplied in $\frac{1}{2}$-, $\frac{5}{8}$-, and $\frac{3}{4}$-inch thicknesses in nominal 6-, 8-, and 10-inch widths. Special patterns are available in 12-inch widths. Drop siding is normally $\frac{3}{4}$ inch in thickness and $5\frac{1}{4}$ inches in face width. (Face width is the coverage width when material is in place.) V-edge siding is $\frac{3}{4}$ inch thick and 6, 8, and 10 inches nominal width (fig. 6). Bevel siding ordinarily is not used for barns, garages, and similar buildings because of cost. Also it is normally laid horizontally over lumber, plywood, or other wood-base sheathing. Some of the thicker grades are occasionally used without sheathing on small garages when the studs are spaced no more than 16 inches on center. Bevel siding is lapped from 1 to $1\frac{1}{2}$ inches depending on spacing required between window heights.

Drop siding (fig. 1-3), because of its uniform thickness, is most often used without sheathing and is applied horizontally. Drop siding has a $\frac{3}{8}$- to $\frac{1}{2}$-inch lap. Matched pattern drop siding is also available. V-edge dressed and matched siding may be applied horizontally or vertically. When it is applied vertically, blocking is required between the studs or vertical members to provide for nailing, if sheathing is not of plywood or lumber. Board and batten combinations require the same type of backing.

The lap of the bevel siding, combined with the actual width, makes it necessary to use from 120 to 150 surface feet for every 100 square feet of surface to be covered. Drop siding requires somewhat less than 115 board feet of 1- by 6-inch drop siding to cover 100 square feet of surface, discounting cutting loss.

Other siding materials that are available in many lumberyards include paper-overlaid plywood siding, and medium- and high-density hardboard. These materials can usually be obtained in 4-foot-wide sheets or prepackaged in narrow sections ready for installation. Medium- or high-density overlaid plywood sheets in ½-inch and greater thicknesses can be applied vertically directly on framing members with proper spacing, serving both as sheathing and siding, for barns and other buildings. Batten strips are normally used over the vertical joints.

Flooring

Flooring (fig.1-3) is made chiefly of hardwood such as maple, oak, birch, and beech, and of the harder softwood species such as Douglas-fir and southern pine. At least one of the softwoods and two of the hardwoods are stocked in most lumberyards. Flooring is usually of nominal 1- by 3- and 1- by 4-inch sizes. Dressed thickness is $25/32$ inch, and face widths are 2¼ and 3¼ inches. Thicker flooring, in Douglas-fir and southern pine, is available and is used for porch flooring without need for a subfloor. Edge grain proves the most satisfactory. Thinner hardwood flooring, usually square-edged, is another form sometimes used as a finish floor.

Block or parquet flooring in 3- or 5-ply plywood or laminated form is available in ½- and $25/32$-inch thicknesses, and is also made in particleboard form. Block flooring is usually installed with a mastic adhesive. When placed on concrete floors, a sealer is used under the block floor and a vapor barrier under the concrete slab to prevent moisture problems.

Edge-grained flooring shrinks less in width than flat-grained flooring, is more uniform in texture, wears more uniformly, and joints do not open up as much. Flat-grained flooring costs less and is commonly used where appearance is not important. Both edge- and flat-grained flooring are carried in stock by many dealers.

Most softwood flooring is either southern yellow pine or Douglas-fir. The higher grades are commonly finished with varnish or sealer. The lower grades are perhaps most suited for dark-stained or painted finishes. Southern pine has five grades: A, B, C, D, and No. 2 flooring; A and B grades are often combined and are classed as B and Better. Douglas-fir has edge- and flat-grained classifications: edge grain in B and Better, C, and D grades, and flat grain in C and Better, D, and E grades.

Hardwood unfinished flooring has many grade classifications depending on the species. Oak flooring has edge-grained grades classed as Clear and Select. Flat-grained oak flooring can be obtained in grades of Clear, Select, No. 1 Common, and No. 2 Common. Maple, birch, and other species have classifications of First, Second, and Third grades.

Exterior Molding and Finish

Exterior moldings and finish are used on cornices, at gable ends, and other finish areas of buildings. Houses may ordinarily be designed with a closed cornice for a desired architectural appearance, but sheds, barns, and other buildings are usually constructed with an open cornice or rafter overhang. Exterior moldings are usually furnished in clear ponderosa pine, southern yellow pine, and Douglas-fir. Many types and sizes are available including crown molding, brick molding, and bed molding as well as moldings used for door and window trim.

Exterior finish material is furnished in the Select grades in nominal sizes from 1 by 2 to 1 by 12 inches and also in $1\frac{1}{4}$ by 6 and $1\frac{1}{4}$ by 8 inches, all surfaced four sides (S4S). The nominal 1-inch finish is used for cornice construction and the 1¼ inch at gable ends or other areas where siding terminates. Woods used include ponderosa pine, western red cedar, redwood, and west coast hemlock.

Shingles

Most wood shingles available in retail lumberyards are of western red cedar, although redwood, white cedar, and cypress are also sometimes stocked. Three grades of shingles are classed under Red Cedar Shingle Bureau rules in three lengths:

No. 1 Blue Label shingles are all clear, all heart, and all edge grain, and are used for the best work as they are less likely to warp. No. 2 Red Label shingles have clear butts about two-thirds to three-quarters of their length and may contain some flat grain and a little sapwood. This grade is often used for roofs of secondary buildings or to cover sidewalls.

No. 3 Black Label shingles have knots and other defects that are undesirable for surface exposure, but have a 6- to 10-inch clear butt depending on their length. This grade is sometimes used as the undercourse in double-course application of sidewalls. An undercoursing shingle is produced expressly for use on double-course sidewalls.

Shingles are produced in three lengths—16, 18, and 24 inches. The 16-inch shingle, the one most likely to be stocked by retail lumberyards, has a standard thickness designated as 5/2-16 (five shingles measure 2 inches thick at the butt when green). The 16-inch shingles are based on a 5-inch exposure when used on roofs, and four bundles will cover 100 square feet (one square). When used in single-course sidewall application, three bundles of 16-inch shingles will cover 100 square feet laid with a 7-inch exposure. Bundled shingles come in random widths of 3 inches and up. Five 18-inch shingles measure 2¼ inches at the butt, and four 24-inch shingles measure 2 inches at the butt when green.

Door and Window Frames

Wood door and window frames, sash, and other similar millwork items are sometimes available in retail lumberyards in standard sizes. Sash and door manufacturers produce ready-hung window units, and the frame, weather-stripped sash, and trim are prefitted, assembled, and ready to be placed in the rough wall opening. However, in smaller retail yards it is usually necessary to order before actual use because many window and door sizes and styles are not stock items.

Ponderosa pine is a species used by most manufacturers for frames and window sash, but southern pine and Douglas-fir are sometimes used for frame parts. Frames for outside doors are usually provided with oak sills to increase their resistance to wear. However, some sills of the softer woods are supplied with metal edgings located at the wearing surfaces.

Most present-day millwork such as door and window frames, sash, and exterior doors are treated at the factory with a water-repellent preservative. This treatment not only aids in resisting moisture but also in minimizing decay hazards.

Plywood

Because plywood is widely available, relatively low in cost, and easy to apply, it can be used to advantage in the construction of homes and farm buildings. It is principally used as a covering material such as subfloor, wall sheathing, and roof sheathing. It is often used for walls and roofs without additional covering for secondary farm buildings. Plywood may also be used for doors of barns and other buildings, and interior lining of barns and milking parlors. It may be used for cabinetwork and as an interior finish wall panel material fabricated in many forms from a variety of species.

The two common types are Interior and Exterior, and these names designate their recommended uses. Sheathing grades are also available. One form of plywood has a resin-impregnated paper overlay on two sides; in this form it is sometimes used as an exterior siding or finish without the benefit of sheathing. This type of plywood is made with waterproof glue and consequently is suitable for exterior use.

Both Exterior and Interior types are available with a variety of sizes and grades of face veneers, ranging from A-A and paper-overlaid faces to C-D sheathing grade. The following are general thicknesses and grades commonly used in frame construction:

Plywood in the Standard interior grade commonly used for wall sheathing should be $3/8$ or $1/2$ inch thick if a siding is applied over the plywood. Rough-textured or patterned exterior plywood (stained finish) used as exterior finish without sheathing is usually $1/2$ inch or more thick depending on the spacing of the studs and the species of plywood. Plywood roof sheathing (Standard interior, C-D) should be at least $3/8$ inch thick if Douglas-fir or southern pine plywood is used and rafters are spaced 16 inches on center. When rafters are spaced 24 inches on center, plywood sheathing should be at least $1/2$ inch thick if Douglas-fir or southern pine is used, and $5/8$ inch thick if other western softwoods are used.

Douglas-fir or southern pine plywood used as subflooring should be at least $1/2$ inch thick and other softwood plywood $5/8$ inch thick when strip flooring is employed. When wood block finish floor is specified for houses, plywood should be $5/8$ inch thick. For a resilient finish floor, the plywood should be $3/4$ inch thick. In floors for sheds and barns, plywood can be 5/8 or 3/4 inch thick depending on the spacing of the joists. Side- and end-matched Douglas-fir plywood in 1 1/8-inch thickness is available for use when supports are spaced as much as 48 inches on center. Plywood used for subfloor and for wall and roof sheathing may be Interior or Exterior. For exterior use, plywood should always be Exterior type.

Hardwood plywood is available in a number of species; perhaps its main use is as a finish covering. The three types available include: Type 1, fully waterproof bond; type 2, water-resistant bond; and type 3, moisture-resistant bond. Details are outlined in U.S. Commercial Standard CS 35. Grades consist of Premium; Good, which is suitable for natural finish; Sound, suitable for a smooth painted surface; Utility, which might be used as sheathing or similar coverages; and Backing. Knots, splits, and other defects are allowed in the Utility and Backing grades.

Much hardwood plywood is used as veneers for flush-solid and hollow-core doors. Because of its variety of uses, standard hardwood plywood is available in widths of 24 to 48 inches and lengths from 48 to 96 inches.

Structural Insulating Board

Many types of sheet materials in addition to plywood are being used for sheathing walls because they are easily applied and resist racking. Structural insulating board sheathing in 1/2- and 25/32-inch thicknesses is available in 2- by 8-foot and 4- by 8-foot sheets. The 2- by 8-foot sheets are applied horizontally and usually have shallow V or tongued-and-grooved edges. The 4- by 8-foot sheets are square-edged and applied vertically with perimeter nailing. These building boards are made water resistant by means of an asphalt coating or by impregnation.

When insulating board sheathing is applied with the 2- by 8-foot sheets horizontally, the construction normally is not rigid enough. Auxiliary bracing, such as 1- by 4-inch let-in bracing, is necessary.

A wall with enough rigidity to withstand wind forces can be built with 4- by 8-foot panels of three types—regular density sheathing $25/32$ inch thick, intermediate density material ½ inch thick, or nail-base grades. Panels must be installed vertically and properly nailed. Each manufacturer of insulating board has recommended nailing schedules to satisfy this requirement.

Interior structural insulating board ½ inch thick and laminated paperboard in ½- and ⅜-inch thickness may be obtained in 4- by 8-foot sheets painted on one side, or in paneled form for use as an interior covering material. These materials are also produced in a tongued-and-grooved ceiling tile in sizes from 12 by 12 inches to 16 by 32 inches; thicknesses vary between ½ and 1 inch. They may be designed to serve as a prefinished decorative insulating tile or to provide acoustical qualities. The present practice of manufacturers is to furnish interior board either plain or acoustical with a flamespread-retardant paint finish.

Medium Hardboard

Medium hardboards are generally available in nominal 7/16- and ½-inch thicknesses in 4-foot-wide sheets or in the form of siding. This material provides good service when used as exterior coverage in sheet form or as lap siding. The 4- by 8-foot sheets are applied vertically, with batten strips placed over the joints and between for decorative effect.

High-Density Hardboard

High-density hardboard in standard or tempered form is commonly supplied in $\frac{1}{8}$- and $\frac{1}{4}$-inch-thick sheets of 4- by 8-foot size. It may be used for both interior and exterior covering material. As with plywood or medium hardboard, the high-density hardboard in the thicker types can be applied vertically with batten strips, or horizontally as a lap siding.

It is often used in the construction of barn doors and for interior lining of barns and other buildings. In perforated form, both types of hardboard are used as soffit material under cornice overhangs to ventilate attic spaces. In untreated form, high-density hardboard of special grade is also used as an underlayment for resilient flooring materials. Hardboards can be obtained with decorative laminated surfaces that provide a pleasing appearance as interior paneling.

Particleboard

Particleboard, a sheet material made up of resin-bonded wood particles, is most often used as an underlayment for resilient flooring. It is also adaptable as covering material for interior walls or other uses where they are not exposed to moisture. Particleboard is usually supplied in 4- by 8-foot sheets and in $\frac{3}{8}$-inch thickness for paneling, in $\frac{5}{8}$-inch thickness for underlayment, and in block form for flooring. It is also used for cabinet and closet doors and as core stock for table tops and other furniture.

Interior Finish and Millwork

Interior finish and millwork include doorjambs and doors; casing, base, base shoe, stool, apron, and other trim and moldings; stair parts; and various cabinets, fireplace mantels, and other manufactured units. Such interior trim as casing and base is stocked in most retail lumberyards in several patterns and at least one species of wood.

Ponderosa pine, Douglas-fir, and southern yellow pine are the softwoods usually available. Oak and birch are hardwoods most likely to be stocked by lumberyards. Species not carried in stock may be obtained and manufactured on special order. Interior trim, moldings, and other interior finish are ordinarily furnished in a clear grade. Paneling in pine, cedar, and similar woods usually contains knots and other grain variations for a decorative effect.

Inside doorjambs are $3/4$ inch thick and vary in width from $4\frac{1}{2}$ to $5\frac{3}{8}$ inches, depending on the type of interior wall finish—drywall or plaster. Base may vary in size from 7/16 by $2\frac{1}{4}$ inches to 9/16 by $3\frac{1}{4}$ inches and wider. The modern trend is toward a narrow base, except in strictly traditional interiors. Base shoe is $1/2$ by $3/4$ inch in size, although quarterround, $3/4$ by $3/4$ inch, is sometimes used as a molding between the base and the finish floor.

Casing and other trim around door and window frames may be obtained in several patterns and sizes. Two common patterns are the "Colonial" type with a molded face and "Ranch" or "Bevel" trim with a simple beveled face and rounded corners. These types of casing are usually $5/8$ or $3/4$ inch thick and $2\frac{1}{4}$ to $2\frac{3}{4}$ inches wide.

Cabinets, fireplace mantels, and stair parts are usually special-order items that the lumber dealer must order from the manufacturer. Interior doors in flush or panel type are often stocked in standard sizes by many of the larger retail yards.

Interior doors are normally $1\frac{3}{8}$ inches thick and vary in width from 2 feet for small closets, to 2 feet 6 inches and wider for other doors. Standard height is 6 feet 8 inches for doors used on the first floor and 6 feet 6 inches for doors used on the second floor. There is a trend in some interior designs of houses to use full wall-height doors. This eliminates the need for headers, head casing and trim, and other construction details associated with lower doors.

Standard exterior doors are 1¾ inches thick and 6 feet 8 inches high. Panel doors with solid wood rail and stiles, and solid-core flush doors are the types most often stocked in retail lumberyards for exterior use.

IMPORTANT POINTS IN CONSTRUCTION AND MAINTENANCE

Wood has withstood the test of time as a building material. On every hand one can see homes and other buildings which prove the permanence of proper wood construction.

Defects that show up in buildings are frequently wrongly ascribed to the type of material used, when they are actually often due to the condition of the material when installed, or to the design, construction, or maintenance of the building. When problems arise, all these factors should be reviewed as possible causes.

Perhaps more unsatisfactory service has resulted from the common failure to use wood at the proper moisture content than from any other cause. The next few pages will outline recommendations for the proper moisture content of wood used in construction. Defects caused by Framing and Finish lumber that is not at the proper moisture content are often difficult or costly to correct.

After the house has been constructed with dry lumber, it is important that the materials remain dry in service. Proper construction details and good carpentry are necessary to assure that moisture does not enter the walls or the interior of the house, for example, proper flashing and clearances (fig.1-4). Excessive changes in its moisture content induce swelling and shrinkage of wood, and are often responsible for plaster cracks and other inconveniences.

How Dry Should Wood Be When Installed?

The installation of wood at the proper dryness means practically no serious shrinkage later. Wood at the time of installation should be seasoned to about the average moisture content that it will have in service. The moisture content of interior trim at the time of installation should be between 5 and 10 percent in most parts of the United States.

In the southern coastal regions, where the humidity is high, the moisture content should be between 8 and 13 percent; for the dry southwestern region, where the humidity is low, the moisture content should be between 4 and 9 percent.

The moisture content of sheathing, framing, siding, and exterior trim at the time of installation should be between 9 and 14 percent in most parts of the United States, and between 7 and 12 percent in the dry southwestern regions.

Determining Moisture Content of Wood

It is very difficult to tell how dry a piece of wood is by looking at it or feeling it. How then can a determination be made? Two means of measurements are available—by use of an electric moisture meter or by the ovendrying method. In addition, there is one way of getting an approximation.

The moisture meter is simple and fast to use and permits determination of moisture content without cutting the board. Several models are available. When used on wood with a moisture content below about 30 percent, these meters can be quite accurate.

The most accurate method of determining moisture content is by ovendrying specimens—

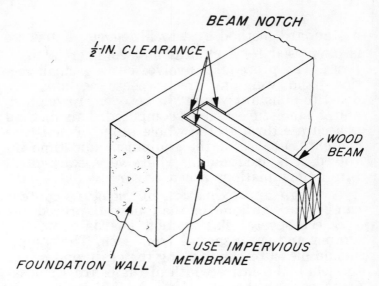
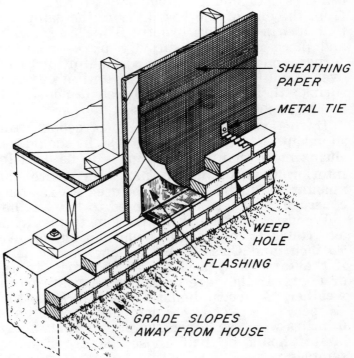

FIGURE 1-4. Proper flashing of masonry veneer and beam notch detail assures that the wood is kept dry.

a standard method by which degree of dryness is expressed for technical and commercial purposes. The procedure involves cutting small sections and weighing them; these sections are dried to constant weight in an oven, reweighed, and the moisture content computed. This method is accurate through the whole range of moisture content. Because of the equipment and time involved, it is used mainly where very exact moisture determinations are necessary.

A rough approximation of moisture content can be made at home by the following procedure:

Select several flat-grained boards from the lumber and cut a sample from each. The sample should measure 1 inch along the grain and be cut to include the entire width of the board (at least 6 inches). It should be cut about 6 to 8 inches from the end of the board. Trim the sample so that it will measure exactly 6 inches in width and place it in a warm, dry place—near the furnace, on a heat duct, on a radiator, or in the oven—and leave it 48 hours or until it ceases to shrink; then measure the 6-inch dimension to determine how much it has shrunk.

If the wood is classed C in freedom from shrinkage (table 1, column 4), it should not shrink more than 1/8 inch if it is to be used for interior trim or finish, nor over twice that amount (1/4 inch) if it is to be used for framing, coverage, or where it is exposed to the weather.

Woods classed as B in freedom from shrinkage (table 1, column 4) should not shrink over 3/32 inch, and Class A woods not over 1/16 inch if they are to be used for interior trim, finish, or floors. If they are to be used exposed to the weather, B woods should not shrink more than 3/16 inch, or A woods 1/8 inch. For lumber under 6 inches wide use 3-inch samples. The shrinkage limits should be half those listed for 6-inch samples.

Edge-grained lumber shrinks only about one-half as much as flat-grained. If it is not possible to obtain a flat-grained sample, an edge-grained sample may be used, but the shrinkage should not be over half that shown for flat grain. It is best not to use edge-grained samples or samples shorter than 6 inches; not only are they more difficult to measure, but they do not give so reliable an indication of the adequacy of seasoning.

Keeping Lumber Dry in Service

Dry wood takes up moisture not only from actual contact with water but from other sources commonly overlooked. It may collect moisture in the form of vapor from damp air or from damp plaster, concrete, soil, or brickwork. Like many other building materials, wood will absorb moisture that has condensed on it, as well as rain or snow that has entered joints and crevices.

The protection of wood from moisture usually requires that it be kept from contact with soil and water; that free circulation of air be provided in damp areas; and that exposed surfaces be protected with paint, varnish, or other coatings. Protective coatings reduce but do not entirely prevent moisture absorption, and therefore should not be relied upon to compensate for poor drainage and poor ventilation. Decay hazards as associated with the moisture content of wood in use are discussed in a later section.

Avoiding contact of wood with moisture is of prime importance in considering construction details. Special care must be used at the grade line of a structure or at any point where moisture might come in. Protection from moisture in the ground should be provided even in temporary or portable buildings.

A little additional care at the start in selection of material and construction details will eliminate the later need for frequent replacement of skids, sills, and framing members (fig. 1-5). The use of treated wood sills or placing the structure on masonry blocking for good air circulation is good practice. A ground cover will minimize movement of water vapor from the ground and prevent the wood from retaining high levels of moisture content. Vapor barriers such as polyethylene, roll roofing, or duplex asphalt papers are satisfactory.

Clearance of wood parts above the finish grade and drainage of water away from the building

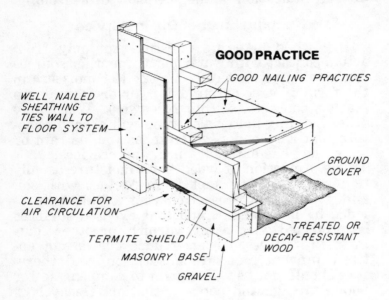

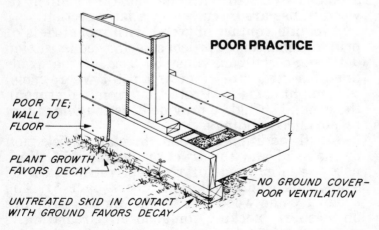

FIGURE 1-5. Good and poor practice for foundations of temporary buildings.

by means of a splash block or tiling are also important factors (fig.1-6). It is difficult to miter siding at corners to prevent moisture en-

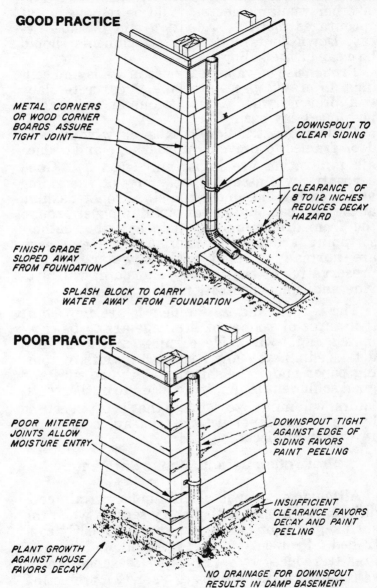

FIGURE 1-6. Good and poor practice with siding joints and downspouts.

try, especially in the wider patterns. Generally, it is better practice to use corner boards or metal corners. Plant growth against the siding or other wood members should be removed as it encourages moisture retention and possible decay. Downspouts and other attachments should be clear of the siding.

Proper use of vapor barriers in walls and ceilings, in crawl spaces, and under concrete slabs will prevent wood from becoming wet and a possible decay hazard.

Correct construction details at window and door frames to prevent rain leakage and reduce air infiltration are important (fig.1-7). Good carpentry will assure tight joints of the siding at the casing and under the sill. Proper flashing at the drip cap, and use of building paper around the framed opening will help as will weather-stripping around the sash. Frames and sash are normally treated with a water-repellent preservative at the factory and paint will provide additional protection.

The cornice and gutter details are important if hazards of poor roof drainage are to be eliminated (fig.1-8). Wide cornices and good drip details eliminate many hazards. A width of roofing paper under the shingles at the cornice and good soffit ventilation, in addition to outlet ventilators, will minimize damage that is often caused by ice dams. Other details relating to moisture in wood are discussed later.

Preventing Defects Due to Shrinkage

Although wood will shrink under certain conditions, it will give satisfactory service when the shrinkage factor is recognized and properly controlled. Problems due to shrinkage can be greatly reduced by: (1) using seasoned woods as required by conditions of use; (2) protecting by paint, water repellents, or other protective coatings all exposed surfaces of dry wood in

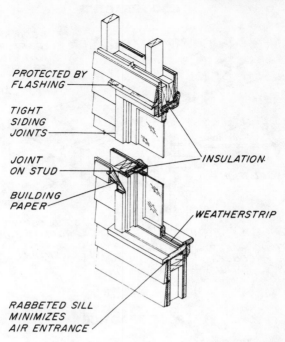
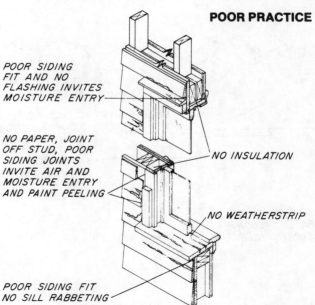

FIGURE 1-7. Good and poor practice with frames.

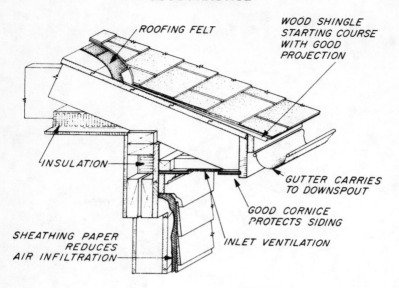
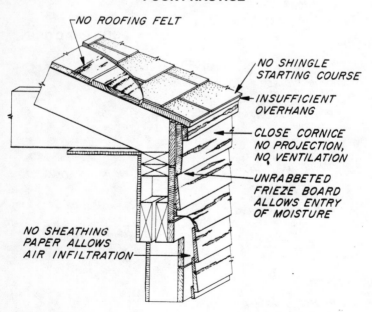

FIGURE 1-8. Good and poor practice in cornice construction.

place so that rapid moisture changes will not occur; (3) selecting woods with low inherent shrinkage (table 1); or (4) using edge-grained material in preference to flat grain for critical uses.

Following the first two rules will insure wood that meets the ordinary requirements of construction. More exacting requirements, such as those of doors, window sash, and frames, require in addition either the selection of woods from the low- or moderate-shrinkage groups or the use of edge-grained material. Special conditions often prevent the application of all four rules. One or more of the rules, however, can always be applied, so as to enable wood to meet the requirements satisfactorily in most cases.

Preventing Decay

The simplest way to prevent wood from decaying is to keep it dry. This means protecting it from common decay hazards caused by leaks, by the contact of wood with the ground, or by contact of wood and water. It also means protecting wood from such commonly unrecognized decay hazards as are caused by small amounts of water that get into the wood and cannot get out. Water is often held in the wood by some type of covering or by lack of ventilation or drainage. Many of these unrecognized decay hazards are at joints that are exposed to the weather and at surfaces where the wood is in contact with other materials. Frequently it is cheaper and easier to change a detail of construction to keep the moisture out than to follow poor design and rely on decay-resistant wood or paint coatings.

Construction Details

Various figures in this bulletin show good and poor practice in several important building details. Four main principles of design to prevent decay of wood members are:

1. Free drainage or good construction details where wood contacts flat areas, such as posts in contact with a concrete slab.

2. Good ventilation to prevent the accumulation of damp air in crawl spaces, under porches and steps, and around the roofing and rafters of barns. Ground covers of roll roofing, polyethylene, or other similar materials will minimize the movement of ground moisture into crawl spaces or under porches. Any enclosed space should be ventilated.

3. Protection of wood from condensation, such as occurs on cold-water pipes and on window glass, especially in dairy barns, bathrooms, kitchens, and all artificially humidified rooms. Insulating cold-water pipes reduces the relative humidity, and hanging them free of joists prevents condensation on the wood. The use of storm windows will usually minimize or eliminate condensation on the glass surfaces.

4. Protective coatings or coverings, such as roofing felt, tarred and mopped down, or polyethylene will prevent absorption from damp concrete, masonry, or earth. The use of a vapor barrier under the concrete slab will eliminate the movement or absorption of ground moisture.

Examples of poor construction practices that provide conditions favorable for decay, and examples of protective measures and good design aimed at reducing decay hazards in wood construction are illustrated in the figures that follow.

At cornices (fig.1-8), good drainage will eliminate the hazards of rain entering the overhang.

Correct method of placing wood posts or columns on a concrete floor is shown in figure 1-9. A concrete pedestal or block to raise the bottom of the post above the floor is an important factor.

A good method of hanging cold-water pipes is illustrated in figure 1-10. When pipes contact the

GOOD PRACTICE

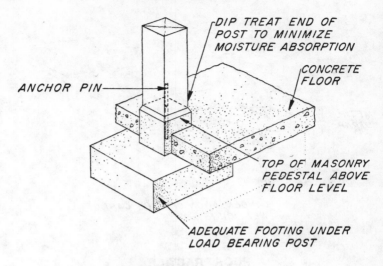

POOR PRACTICE

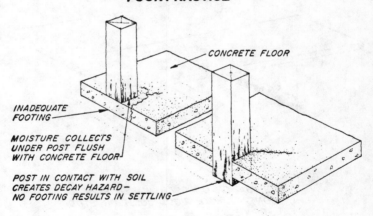

Figure 1-9. Good and poor practice with wood posts.

joists, condensation on the pipes during the warmer months keeps the joist wet and invites decay. Hanging these pipes free of joists avoids this problem. Covering the cold-water pipes with insulation will also help.

Good practice in the manufacture and use of window sash is shown in figure 1-11. Priming and

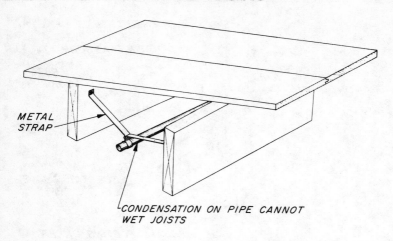
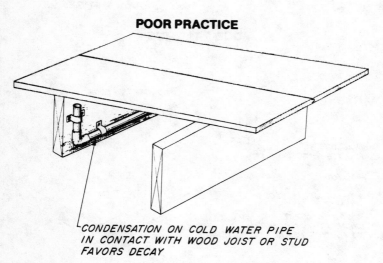

Figure 1-10. Good and poor practice for supporting water pipes.

back puttying are important in preventing moisture entry behind the putty. A water-repellent preservative dip will also aid in reducing moisture entry.

Examples of both good and poor construction practices in placing wood floors over concrete

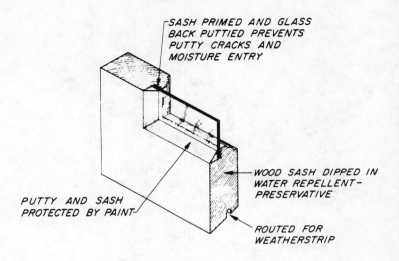

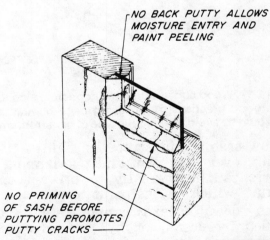

FIGURE 1-11. Good and poor practice with window glass.

slabs are given in figure 1-12. A good vapor barrier under the concrete slab is an important factor in preventing moisture entry.

Examples of both good and poor construction practices in installing drip cap over exterior of

GOOD PRACTICE

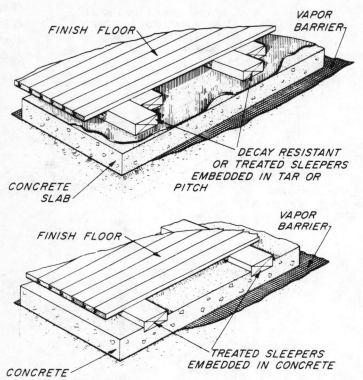

Figure 1-12. Wood construction over concrete slab. Left page shows good practices; right facing page depicts practices that should be avoided with this type of construction.

window sash are illustrated in figure 1-12. Use of flashing over drip cap, and overlapping siding on drip cap help greatly to prevent entry of moisture which causes paint to peel.

Many decay hazards cannot be eliminated or modified by design or by protective coatings. The conditions of use may be such that wood must touch the ground or water. For example, there is no practical method by which wood piers to buildings, fenceposts, sills on the ground, or sleepers embedded in concrete can be kept dry. Protection against decay in such uses lies in preservative treatment. One type of use that

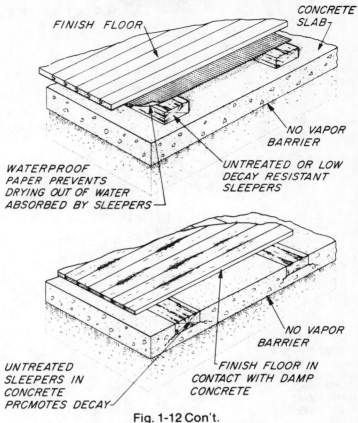

Fig. 1-12 Con't.

requires treated or decay-resistant woods is shown in figure 1-12.

Preservative-Treated Material

Good preservative treatment will insure longer life for wood that is not resistant to decay but must be used in contact with the ground or in a moist area. A normal life of 3 years for an untreated southern pine fencepost can be increased to 35 years or more if the post is effectively treated. Wood, when effectively treated, can be considered on the same basis as other so-called permanent materials.

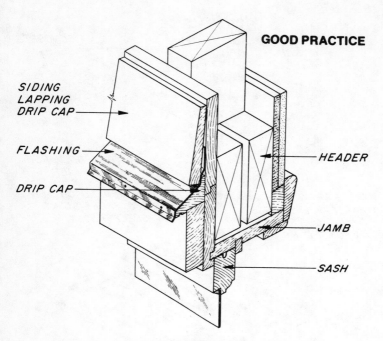
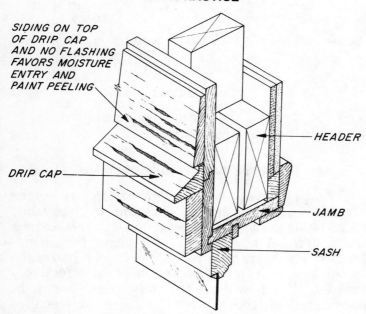

FIGURE 1-13. Good and poor practice at window drip cap.

Wood-preserving methods are of two general types: (1) Pressure processes in which the wood is impregnated in closed vessels under pressures considerably above atmospheric pressure; and (2) nonpressure processes including soaking, diffusion, brushing, spraying, and dipping. Pressure processes generally provide greater protection than nonpressure processes.

Preservatives for wood are also of two types: (1) Oils, such as creosote, creosote solutions, and pentachlorophenol, and copper naphthenate in oil carriers; and (2) waterborne salts applied as water solutions.

Preservative oils are ordinarily used where resistance to leaching is desired, such as fenceposts, exposed poles, and timbers in contact with the ground. However, preservative oils applied by thorough methods may be objectionable because of odor, and because of their effect on the cleanliness, combustibility, and paintability of wood. Water-solution preservatives are used principally for treating wood that is not to be in contact with the ground or water, and where the treated wood requires painting.

Water-repellent preservatives have been used for window sash and frames for a number of years, and now are also being used on siding and exterior trim where water repellency and some protection is advantageous.

For most home and farm uses that require good preservative treatment, it is advisable that the treatment be in accordance with a recognized standard such as Federal Specification TT-W-571. Retentions recommended for preservative oils vary from 6 to 10 pounds per cubic foot and from 0.3 to 1.0 pound per cubic foot for waterborne preservatives.

The hot and cold bath is perhaps the most effective treatment that can be applied by the user even though it is not as satisfactory as pressure treating.

If the wood is to be in service above the ground, as in porches, steps supported on masonry, or railings, considerable protection from decay can be gained by soaking the lumber for about 15 minutes in the cold preservative. This should be done after the wood has been machined and drilled. A suitable preservative that will permit the wood to be painted later is a 5 percent solution of pentachlorophenol, plus water repellents, in light oil. This preservative is commonly available at lumber dealers.

How and When to Paint

The purpose of painting or staining exterior wood is to improve and maintain appearance. Painting prevents the wood from weathering and reduces cracking and warping; thereby it prevents the appearance of age and neglect that even later painting cannot remove. Paint does *not* prevent decay.

Painting characteristics of different kinds of wood are indicated in table 1. Edge-grained surfaces of all species of wood are superior to flat-grained surfaces in their ability to hold paint. Carefully selected edge-grained boards of redwood and western red cedar are commonly used for house siding, but flat-grained boards are generally used in most farm structures.

Because of high swelling and poor paint-holding properties, flat-grained boards are frequently better finished by staining with a red or brown pigmented penetrating stain. This kind of a finish penetrates into the wood without forming a continuous film on the surface. Therefore, it will not blister, crack, or peel, even if excessive moisture penetrates into the wood.

One way to improve the performance of stained flat-grained surfaces is to leave the wood roughsawn. Allowing the wood surface to weather several months also roughens the surface and improves it for staining. Penetrating stains are effective finishes for weathered exterior plywood. Surfaces that are rough textured take up more stain, and insure a more durable stain.

In painting wood a few simple, tried-and-tested procedures will result in a coating that will give many years of satisfactory service:

Step 1—Water-repellent preservative.—Treat with water-repellent preservative before painting to protect wood against the entrance of rain or heavy dew. It is especially important that window sash and trim be so treated. This protection can be applied in two ways:

    a. Use material treated by the manufacturer. On the job, retreat any cut ends by brushing on the solution.

    b. Apply the entire treatment on the job by brushing. Be careful to brush the preservative well into lap and butt joints. Allow 2 warm, sunny days for adequate drying of the preservative before painting.

Step 2—Priming.—For the first or prime coat on wood, and for spot priming when repainting, use a linseed oil-base paint free of zinc-containing pigments. Follow the spreading rates recommended by the manufacturer; do not spread paint too thin. You should not be able to see the grain of the wood after priming. Priming with this oil-base paint is necessary even when the second coat is to be an exterior emulsion or latex paint.

Step 3—Finish coat.—For best results:

    a. Use a high-quality paint.

    b. Apply two top coats of oil-base or exterior latex (vinyl or acrylic) paint. A two-coat (primer and one top coat) job of low-quality paint may last only 3 years, but a three-coat job with good-quality paint may last 10 years.

    c. To avoid intercoat peeling of paint, apply top coats within 2 weeks after the primer. Do not prime in the fall and then delay top coats until spring. Instead, treat with water-repellent preservative immediately and delay *all* painting until spring.

d. To avoid temperature blistering, do not apply oil-base paints on a cool surface that will be heated by the sun within a few hours. Rather, paint on the side the sun is on.

e. To reduce wrinkling and flatting of oil-base paint, and water marks on latex paint, do not paint late in the evenings of cool spring and fall days when heavy dews frequently form.

f. In areas where mildew is a problem, use oil-base paints that contain zinc oxide pigment, or latex paints with fungicide for top coats.

Step 4—Repainting.—A new paint is only as good as the old paint beneath, so consider these general rules:

a. Before repainting, wash old glossy and unweathered surfaces, or roughen them with steel wool to remove contaminants that may interfere with adhesion to the next coat.

b. Repaint only when the old paint has weathered so it no longer covers or protects the wood. Where wood surfaces are exposed, remove loose paint and spot prime with the zinc-free paint primer.

c. For the top coats, use good-quality latex or oil-base paint that is known to give good service.

Occasionally oil-base paint applied to wood that is very wet or green may blister, but wet wood is seldom the cause. Paint blistering is more often caused by water that works into and accumulates in the sidewalls of structures, back of the painted siding. Cold-weather condensation, rainwater, and ice dams formed on roofs by melting snow are major sources of moisture responsible for paint problems.

Preventing Entrance of Air

Infiltration of air often results in a cold and uncomfortable house even when the best kind and grade of lumber are used. Good tight construction is obtained principally by good workmanship and by use of dry lumber, good building paper properly applied when wood sheathing is used, and woods classed in table 1 as A or B in freedom from shrinkage and warping.

Poor workmanship allows the entrance of air and water because of poor fitting of miter joints at corners (fig. 1-6). Failure to break joints on studs and to fit siding to window sills also lets air in (fig. 1-7). Building paper well lapped over the wood sheathing practically seals a house against air seeping through the walls, but it cannot entirely protect the house against poor fitting at the openings.

Good building paper most efficiently excludes air when clamped between two coverings, such as sheathing and drop or bevel siding. Plywood, other wood-based panel materials, or similar types of material used in large sheets as sheathing ordinarily do not require building paper to provide a tight wall. However, it is good practice to use 12-inch-wide strips of building paper around door and window frames over sheet materials used for sheathing. Thus, after the application of the siding, air infiltration will be reduced to a minimum.

THERMAL INSULATION CONSIDERATIONS

An important development in modern construction practices is the increasing use of thermal insulation in houses, particularly in dwellings of intermediate and low cost. The basic objective is year-round comfort, since insulation keeps the house warmer in winter and cooler in summer. However, the saving in fuel to heat an insulated house may also justify the added cost of insulation.

Most materials used in construction offer some resistance to the transmission of heat, but wood and wood-based materials have better insulating qualities than many others. In this respect, 1 inch of Douglas-fir is equal to about 12 inches of concrete or stone in resistance to heat transmission. On the other hand, it would take about 2 inches of the wood to equal the insulating qualities of 1 inch of fiberboard.

Classes of Insulating Materials

Materials that have a relatively high resistance to heat transmission are called "thermal insulators" or, more commonly, insulation. Insulating materials may be grouped into the following general classes: A, rigid insulation; B, flexible insulation; C, fill insulation; and D, reflective insulation.

Rigid insulation.—One of the most commonly used types of rigid insulation is structural insulating board. Such boards often combine insulation ability with moderate strength. In this form they are used as sheathing material and as interior finish, in laminated plank form for roof deck covering, or as an acoustic tile. These types are much better insulators on an equivalent thickness basis than solid wood, but are inferior to most flexible insulations.

Flexible insulation.—Flexible materials are ordinarily used only for insulation and are normally manufactured in blanket or batt form. In this form the insulating value is more than three times that of wood of equal thickness. Blanket and batt insulation are treated to resist fire, vermin, and decay. They are generally supplied with a vapor barrier on one side to minimize movement of water vapor through walls and ceilings.

Fill insulations.—Loose, fill-type insulations are made of materials used in bulk form and are poured or blown in place. They are used to fill stud spaces in walls and between joists in attics. Among the products most commonly used are wood fibers, granulated cork, shredded redwood bark. and ground newsprint. Fill materials have insulating values slightly less than the same material in blanket or batt form.

Reflective insulation.—Another class of insulation consists of the materials that reflect radiant heat to a high degree. Reflective insulation is effective only if the reflective surface is adjacent to a free air space of at least $3/4$ inch. For reflective insulation, high reflectivity is required, as provided by aluminum foil or polished metal surfaces or coatings. As an example, aluminum foil is often mounted on the back of gypsum board and thus serves both as a vapor barrier and as reflective insulation.

Where to Insulate

Insulation should be used in ceilings, walls, and floors where wide temperature differences occur on opposite sides of those surfaces. For example, in unheated attics, the insulation should be placed above the ceiling of the rooms below. In heated attics or $1\frac{1}{2}$-story houses, however, the insulation should be located in the attic ceiling and down the slope of the roof to the plate, or to and through the knee wall.

In flat and pitched roofs of frame buildings, where the insulation is placed above the ceiling, a space must be left between the insulation and the roof sheathing for air circulation.

All exterior walls should be insulated including walls between heated rooms and unheated garages, porches, and similar spaces. Floors over unheated crawl spaces or porches should also be insulated.

Besides the benefit of improved living conditions in both winter and summer, added insulation will result in substantial savings in fuel. The trend has been toward greater amounts of insulation, especially in attic areas. Four inches of ceiling insulation was considered more than sufficient at one time but, with increased fuel costs, 6 and 8 inches are now used in many northern areas. Two inches or more are commonly used in the walls in these areas.

Home and Building Plans Available

The Cooperative Farm Building Plan Exchange and the Midwest Plan Service prepare complete working drawings for a large variety of homes and farm buildings adapted to all sections of the country, and all types of agriculture.

Regional committees of housing and farm building specialists in research and extension, guide, advise, and act in the execution of working drawings that are distributed through the State Extension Services at land-grant universities.

In many States, local county agricultural agents can advise about the plans that are available. In other States, the farm building specialist, an extension agricultural engineer at the State Agricultural College, maintains lists, files, and distribution facilities for mailing working drawings. Some States make a small charge to cover the costs of printing and mailing the plans.

A short selected list of some of these home and farm building plans is given in "List No. 5, Popular Publications for the Farmer, Suburbanite, Homemaker, Consumer." This list is available free from the Office of Information, U.S. Department of Agriculture, Washington, D.C. 20250. From this list, individual sheets illustrating and summarizing the main features of each plan may be ordered free These sheets give a brief description of the plans but do not generally give details required for construction. The details are given in the working drawings.

2

HOW TO CONSTRUCT YOUR LOW-COST HOME

In the building of a low-cost home, the matter of cost must be considered at every step—in design, selection of materials, and in construction. Each square foot of area added to a plan increases the cost substantially. It is often wise to omit some features, even though desirable, and to add these at some opportune future time. Selection of satisfactory alternate materials can also account for substantial savings.

THE BASICS

The unit cost of framing and enclosing a basic wood house does not vary a great deal and is generally determined by square footage. However, the type of foundation, materials used on the inside and outside of the basic wood frame, the type of windows and doors, number of kitchen cabinets and amount of other millwork materials, floor covering, and the caliber of the utilities included can vary a great deal and generally govern the overall cost of the house. For example, in an average single-story house, kitchen cabinets, interior doors and trim, and hardwood floors account for about 15 percent of the total cost, and substitute materials or deletions in these areas result in substantial savings. The plumbing, electrical, and heating installations also account for about 15 percent of the total house cost, and savings can also be made by eliminating or delaying some of these phases of construction until later.

The cost of a basement in an average house can also amount to 15 percent of the total cost. It seems justified, therefore, in instances where cost is important, to eliminate a basement and have a crawl space consisting of a foundation of treated wood posts or masonry piers. In a small house, this could result in a saving of up to 50 percent of the cost of a conventional foundation.

With a good plan and adequate construction details, a small house can be constructed at a reasonable cost and yet provide for good family living and be as pleasing in appearance as higher cost houses.

MAJOR HOUSE PARTS

Fig. 2-1 is an exploded view of a single-story, wood-frame house showing the major parts. The floor system, interior and exterior walls, and the roof are the major components of such a house. Houses with flat or low-sloped roofs are usually variations of these systems. The A-frame, post and beam, and pole-frame are other systems used in house construction.

Floor Systems

Fig 2-1 shows a floor system constructed over a *crawl space* area. Supporting *beams* are fastened to treated posts embedded in the soil or to masonry *piers*. In the South, Central, and Coastal areas, provisions must be made for protection from termites. Construction of this type of support for the floor joists has a great advantage because grading is not required and thus it can be used on relatively steep or uneven slopes. Floor *joists* are fastened to these beams and the *subfloor* nailed to the joists. This results in a level, sturdy platform upon which the rest of the house is constructed.

Exterior Walls

Exterior walls, often assembled flat on the subfloor and raised in "tilt-up" fashion, are fastened to the perimeter of the floor platform. Exterior coverings and window and door units are included after walls are plumbed and braced.

Interior Walls

Interior walls are usually the next components to be erected unless *trussed* rafters (roof trusses) are used. Trussed rafters are designed to span from one exterior sidewall to the other and do not require support from interior *partitions*. This allows partitions to be placed as required for room dividers. When ceiling joists and rafters are used, a *bearing partition* near the center of the width is necessary.

Roof Trusses or Roof Framing

Several systems can be used to provide a roof over the house. One consists of normal ceiling joists and rafters which require some type of load-bearing wall between the sidewalls (fig. 2-2A). Another is the trussed rafter system (figs. 2-1 and 2-2B) (commonly called *trusses*). This design requires no load-bearing walls between the sidewalls. A third design consists of thick wood roof decking (fig. 2-2C). A fourth is open beams and decking which span between the exterior walls and a center wall or ridge beam (fig. 2-2D). The truss and the conventional joist-and-rafter construction require some type of finish for the ceiling. The decking (fig. 2-2C) or the beam and decking (fig. 2-2D) combinations can serve both as interior finish and as a surface to apply the roofing material.

MATERIAL SELECTION

There are hundreds of materials on the market which can be used somewhere in the construction of a house. Many are costly and are meant pri-

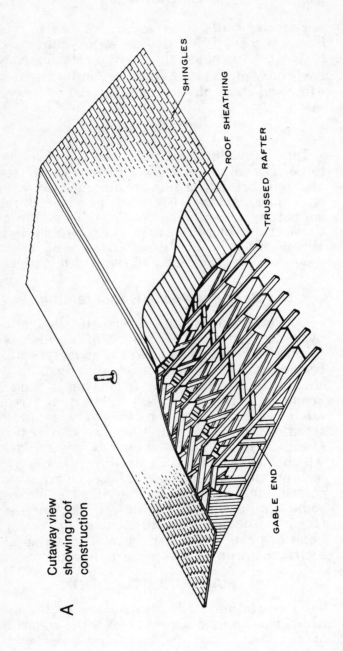

A Cutaway view showing roof construction

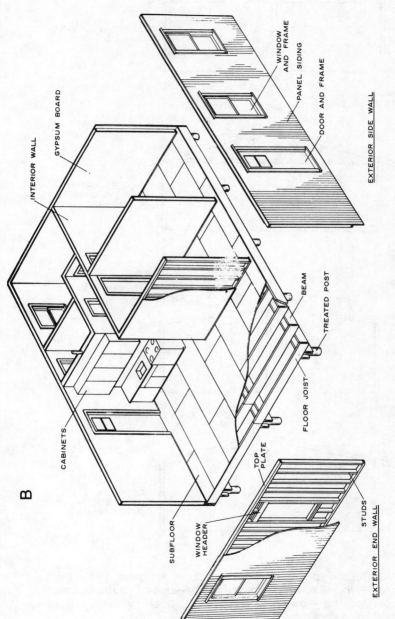

Figure 2-1. Exploded view of wood-frame house.

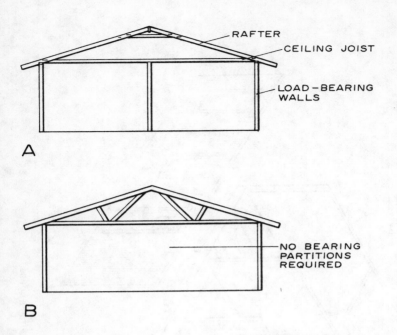
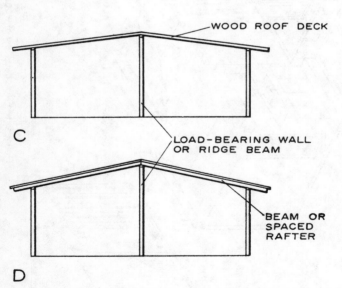

FIGURE 2-2. Types of roof construction. A, Rafter and ceiling joists—sloped roofs; B, trussed rafter—sloped roofs; C, wood roof decking—low-sloped roofs; D, beam with wood or fiberboard decking.

marily for use in the most expensive homes. Others may not be suitable for all of the intended uses. However, among these building materials are many that are reasonable in cost and perform efficiently. Most manufacturers recommend particular uses and application methods for each of their products, and few problems will occur if such recommendations are followed.

Wood

Wood in its various forms is perhaps the most common and well-known material used in house construction. It is used for framing of floors, walls, and roofs. It is sometimes used in board form as a covering material, but more often the covering materials take the form of plywood or other *panel* wood products. Wood is also used as *siding* or exterior covering, as interior covering, as interior and exterior trim, as flooring, in the many forms and types of millwork, and also as shingles to cover roofs and sidewalls.

Wood is easy to form, saw, nail, and fit; even with simple handtools, and with proper use and protection will give excellent service. The *moisture content of wood* used in various parts of a house is important, and recommended moisture contents will be outlined in later sections of this handbook.

There are a number of basic standard wood and wood products used in the *construction* of wood-frame houses. The selection in the proper type and grade for each use is important. The materials can be divided into groups by their use in the construction of a house: some require good strength, others workability, and still others necessitate good appearance.

Treated Posts

Wood posts or poles which are embedded in the soil and used for support of the house should be pressure treated. A number of species are used for these round members. The pressure treatment should conform to Federal Specification TT–W–

571. Pressure treatments normally utilize the oil type of *preservatives* (empty-cell process) or leach-resistant waterborne salt preservatives (full-cell process.)

Dimension Material

Surfaced *dimension* material wood members 2 to 4 inches thick and other wood parts are not full size as they are received from the *lumber yard*. For example, a nominal 2 by 4 may have a finished thickness of 1½ to 1 9/16 inches and a width of 3½ to 3 9/16 inches, depending on the moisture content. These materials are sawn from green logs and must be surfaced as well as dried to a usable moisture content. These processes account for the difference in size between a finished dry member and a rough green member.

The green-dry size relationship should hold for dimension lumber up to 4 inches in nominal thickness.

The first materials to be used, after the foundation is in place, are the floor *joists* and *beams* upon which the joists rest. These require adequate strength in bending and moderate stiffness. The sizes used depend on a number of factors; the *span*, spacing, species, and grade. Recommended sizes are listed in most working plans. The second grade of a species, such as southern pine, western hemlock, or Douglas-fir, is commonly selected for these uses. In lower cost houses, the third grade is usually acceptable.

For best performance, the *moisture content* of most dimension materials should not exceed 19 percent.

Wall *studs* (the structural members making up the wall framing) are usually nominal 2 by 4 inches in size and spaced 16 or 24 inches apart. Their strength and stiffness are not as important as for the floor joists, and in low-cost houses the third grade of a species such as Douglas-fir or southern pine is satisfactory. Slightly higher grades of other species, such as white fir, eastern white pine, spruce, the western white pines, and others, are normally used.

Members used for trusses, rafters, beams, and ceiling joists in the roof framing have about the same requirements as those listed for floor joists. For low-cost houses, the second grade can be used for trusses if the additional strength reduces the amount of material required.

Covering Materials

Floor *sheathing* (subfloor) consists of board *lumber* or *plywood*. Here, too, the spacing of the joists, the species of boards or plywood, and the intended use determine thickness of the subfloor. A single layer may serve both as subfloor and top surface material; for example, 5/8-inch or thicker tongued-and-groved plywood of Douglas-fir, southern pine, or other species in a slightly greater thickness can be used when joists are spaced no more than 20 inches on center. While Douglas-fir and southern pine plywoods are perhaps the most common, other species are equally adaptable for floor, wall, and roof coverings. The "Identification Index" system of marking each sheet of plywood provides the allowable rafter or roof truss and floor joist spacing for each thickness of a standard grade suitable for this purpose. A nominal 1-inch board subfloor normally requires a top covering of some type.

Roof sheathing, like the subfloor, most commonly consists of plywood or board lumber. Where exposed wood beams spaced 2 to 4 feet apart are used for low-pitched roofs, for example, wood decking, fiberboard roof deck, or composition materials in 1- to 3-inch thicknesses might be used. The thickness varies with the spacing of the supporting rafters or beams. These sheathing materials often serve as an interior finish as well as a base for roofing.

Wall sheathing, if used with a siding or secondary covering material, can consist of plywood, lumber, structural insulating board, or gypsum board. The type and method of sheathing application normally determine whether *corner bracing* is required in the wall. When 4- by 8-foot sheets of $^{25}/_{32}$-inch regular or ½-inch-thick, medium-density, insulating fiberboard or $^{5}/_{16}$-inch or thicker plywood are used vertically with proper nailing all around the edge, no bracing is required for the rigidity and strength needed to resist windstorms. There are plywood materials available with grooved or roughened surfaces which serve both as sheathing and finishing materials. Horizontal application of plywood, insulating fiberboard, lumber, and other materials usually requires some type of diagonal *brace* for rigidity and strength.

Exterior Trim

Some exterior trim, such as *facia* boards at cornices or gable-end overhangs, is placed before the roofing is applied. Using only those materials necessary to provide good utility and satisfactory appearance results in a cost saving. These trim materials are usually wood and, if relatively clear of *knots*, can be painted without problems. Lower grade boards with a rough-sawn surface can be stained.

Roofing

One of the lowest cost roofing materials which provides satisfactory service for sloped roofs is mineral-surfaced *asphalt roll roofing*. Asphalt shingles also give good service. Both are available in a number of colors. The material cost of asphalt *shingles* is about twice that of surfaced roll roofing. In a small, 24- by 32-foot house, use of surfaced roll roofing may mean a saving of $40 for material and about $20 for labor, which is less than 1 percent of the total cost of the house. However, an asphalt shingle would normally last longer and have a better apperance than the roll roofing.

Wood shingles have a pleasing appearance for sloped roofs. Although they are usually more costly than composition roofing, they could be used where availability, cost, and application conditions were favorable.

Window and Door Frames

Double-hung, casement, or awning wood windows normally consist of prefitted sash in assembled frames ready for installation. A double-hung window is one in which the upper and lower sash slide vertically past each other. A *casement sash* is hinged at the side and swings in or out. An awning window is hinged at the top and swings out. Separate sash in $1\frac{1}{8}$- or $1\frac{3}{8}$-inch thickness can be used, but some type of frame must be made that includes *jambs*, stops, *sill*, casing, and the necessary hardware. A low-cost, factory-built unit which requires only fastening in place may be the most economical. A fixed sash or a large window glass can be fastened by stops to a prepared frame and generally costs less than a movable-type window. It is normally more economical to use one larger window unit than two smaller ones. Screens should ordinarily be supplied for all operable windows and for doors. In the colder climates, storm windows and storm or *combination doors* are also desirable. Combination units, with screen and storm inserts, are commonly used.

Exterior Coverings

Exterior coverings such as horizontal wood *siding*, vertical boards, boards and *battens*, and similar forms of siding usually require some type of backing in the form of *sheathing* or nailers between studs. In mild climates, nominal 1-inch and thicker sidings are often used over a waterproof paper applied directly to the braced stud wall. There are many sidings of this type on the market in both wood and nonwood materials.

Combination sheathing-siding materials (panel siding) usually consist of 4-foot-wide sheets of plywood, exterior particleboard, or hardboard. Applied vertically before installation of window and door frames, such materials serve very well for exteriors. Plywood may be stained or painted, and the other materials should be painted. Paper-overlaid plywood also serves as a dual-purpose exterior covering material and takes paint well. Wood shingles and *shakes* and similar materials normally require a solid backing or spaced boards of some type.

Insulation

Most houses, even those of lowest cost, should have some type of *insulation* to resist the cold and to increase comfort during hot weather. There are various types of insulation, from insulating fiberboard to fill types, which can be used in the construction of a house. Perhaps the most common *thermal insulations* are the flexible (blanket and batt) and the fill types

A blanket insulation might be used between the floor joists or studs. Batt insulation of various types might be used between floor joists or in the ceiling areas. Most flexible insulations are supplied with a vapor barrier which resists movement of water vapor through the wall and minimizes *condensation* problems. A friction-type batt insulation is also available for use in floors, walls, or ceilings. Fill-type insulation is most commonly used in attic-ceiling areas.

The structural insulating board often serves as sheathing in the wall or as a fair insulating material under a plywood floor. Each material has its place, and selection should be based on climate as well as on cost and utility.

Interior Coverings

Many *dry-wall* (unplastered) interior coverings are available, from gypsum board to prefinished plywood. Perhaps the most economical are the gypsum board products. They are normally applied vertically in 4- by 8-foot sheets or horizontally in room-length sheets with the joint at midwall heights. They are also used for ceilings. Thicknesses range from 3/8 to 5/8 inch. *Butt joints* and corners require the use of tape and joint compound, or a *corner bead*, and add somewhat to labor costs over prefinished materials. Plastic-covered gypsum board is also available at additional cost, but must usually be installed with an adhesive. Hardboards, insulation board, plywood, and other sheet materials are available, as are wood and fiberboard paneling. The choice must be based on overall cost of material and labor as well as on ease of maintenance. Prefinished ceiling tile in 12- by 12-inch sizes and larger is also available.

Interior Finish and Millwork

Interior finish and *millwork* consist of doors and door frames, *base moldings*, window and door *trim*, kitchen and other *cabinets*, flooring, and similar items. The type and grade selected determine the cost to a great extent. Selection of simple *moldings*, lower cost species for jambs and other wood members, simple kitchen shelving, low-cost floor coverings, and the elimination of doors where practical will often make a difference of hundreds of dollars in the total cost of the house.

The most commonly used doors are the flush-type and the panel-type. The flush-type consists of thin plywood or similar facings with a solid or hollow core. The panel door consists of solid side stiles and cross*rails* with plywood or other panel fillers. For exterior types, both may be supplied with openings for glass. Exterior doors are usually 1¾ inches thick and interior doors 1⅜ inches thick.

Door jambs, casings, moldings, and similar millwork of a number of wood species can be obtained. Select the lower cost materials, yet those that will still give good service. Some species in this class are the pines, the spruces, and Douglas-fir.

TABEL 2-1. Recommended schedule for nailing the framing and sheathing of a well-constructed wood-frame house.

Joining	Nailing method	Nails Number	Nails Size	Nails Placement
Header to joist	End-nail	3	16d	
Joist to sill or girder	Toenail	2-3	10d or 8d	
Header and stringer joist to sill	Toenail		10d	16 inches on center.
Bridging to joist	Toenail each end	2	8d	At each joist.
Ledger strip to beam, 2 inches thick		3	16d	
Subfloor, boards:				
1 by 6 inches and smaller		2	8d	To each joist.
1 by 8 inches		3	8d	To each joist.
Subfloor, plywood:				
At edges			8d	6 inches on center.
At intermediate joists			8d	8 inches on center.
Subfloor (2 by 6 inches, T&G) to joist or girder	Blind-nail (casing) and face-nail	2	16d	
Soleplate to stud, horizontal assembly	End-nail	2	16d	
Top plate to stud	End-nail	2	16d	
Stud to soleplate	Toenail	4	8d	
Soleplate to joist or blocking	*Face-nail*		16d	16 inches on center.
Doubled studs	Face-nail, stagger		10d	16 inches on center.
End stud of intersecting wall to exterior wall stud	Face-nail		16d	16 inches on center.
Upper top plate to lower top plate	Face-nail		16d	16 inches on center.
Upper top plate, laps and intersections	Face-nail	2	16d	
Continous header, 2 pieces, each edge			12d	12 inches on center.
Ceiling joist to top wall plates	Toenail	3	8d	

		Number	Size	
Ceiling joist laps at partition	Face-nail	4	16d	
Rafter to top plate	Toenail	2	8d	
Rafter to ceiling joist	Face-nail	5	10d	
Rafter to valley or hip rafter	Toenail	3	10d	
Ridge board to rafter	End-nail	3	10d	
Rafter to rafter through ridge board	Toenail	4	8d	
	Edge-nail	1	10d	
Collar beam to rafter:				
2-inch member	Face-nail	2	12d	
1-inch member	Face-nail	3	8d	
1-inch diagonal let-in brace to each stud and plate (4 nails at top)	Face-nail	2	8d	
Built-up corner studs:				
Studs to blocking	Face-nail	2	10d	Each side.
Intersecting stud to corner studs	Face-nail		16d	12 inches on center.
Built-up girders and beams, 3 or more members	Face-nail		20d	32 inches on center, each side.
Wall sheathing:				
1 by 8 inches or less, horizontal	Face-nail	2	8d	At each stud.
1 by 6 inches or greater, diagonal	Face-nail	3	8d	At each stud.
Wall sheathing, vertically applied plywood:				
3/8 inch and less thick	Face-nail		6d	} 6-inch edge.
1/2 inch and over thick	Face-nail		8d	} 12-inch intermediate.
Wall sheathing, vertically applied fiberboard:				
1/2 inch thick	Face-nail			1½-inch roofing nail.[1]
25/32 inch thick	Face-nail			1¾-inch roofing nail.[1]
Roof sheathing, boards, 4-, 6-, 8-inch width	Face-nail	2	8d	At each rafter.
Roof sheathing plywood:				
3/8 inch and less thick	Face-nail		6d	} 6-inch edge and 12-inch intermediate.
1/2 inch and over thick	Face-nail		8d	

[1] 3-inch edge and 6-inch intermediate.

Factory-built kitchen cabinets are expensive and can cost several hundred dollars in a moderate-size house. The use of open shelving which can be curtained and a good counter is almost a must in a low-cost house. Doors can be added at a later date.

Wood strip flooring or wood tile of hardwood species might be too costly to consider in the original construction, but could be installed at a future time. Softwood floorings or the lower cost hardwood floorings might be within the original budget. The use of lower cost asphalt tile or even a painted finish may be the best initial choice. However, when a woodboard subfloor is used, some type of underlayment is required under the tile. Particleboard, hardboard, and plywood are the most common materials for this use.

Nails and Nailing

In a wood-frame house, nailing is the most common method of fastening the various parts together. Nailing should be done correctly because even the highest grade member often does not serve its purpose without proper nailing. Thus, it is well to follow established rules in nailing the various wood members together. While most of the nailing will be described in future sections, table 1 lists recommended practices used for framing and application of covering materials. Figure 3 shows the sizes of common nails. Most finish and siding nails have the same equivalent lengths. For example, an eight*penny* common nail is the same length as an eightpenny galvanized siding nail, but not necessarily the same diameter.

Painting and Finishing

There are many satisfactory *paints* and finishes for exterior use. *Pigmented stain* is one of the easiest types to apply and is also long lasting. It is available in many colors from light to dark and is generally one of the best finishes for rough or sawn wood surfaces. Exterior paints used on smooth-surfaced siding and trim or on the trim as an accent for stained walls should be applied in several coats for best service. The first may consist of a nonporous linseed oil *primer*. Following coats can consist of latex, alkyd, or oil-base exterior paints. A *water-repellent preservative* provides a natural clear finish for wood surfaces.

Many interior paints are suitable for walls and ceilings. Latex and alkyd types are perhaps the most common, but the oil types are also suitable.

Floor and *deck paints* provide long wearing surfaces. One of the most common finishes for wood floors is the floor *sealer*, which provides a natural transparent surface.

Chimney

Some type of chimney will be required for the heating unit. A masonry chimney requires a rigid concrete base, bricks or other masonry, *flue lining*, plus labor to erect it. A manufactured chimney which is supported by the ceiling joists or rafters may be the best choice from the standpoint of overall cost.

Utilities

Cost of the heating, plumbing, and wiring phases of house construction is usually a high percentage of total cost. These costs can be reduced to a great extent by careful selection and planning. Electricity is readily available in many areas and should be included in most low-cost houses. A minimum number of circuits and few switched outlets will aid in reducing the overall installation cost.

Heating units might consist of a low-cost space heater for wood, oil, or gas, or a small, central forced-air system with a minimum amount of duct work. The difference between the two may amount to several hundred dollars. The space heater may be more than sufficient for houses constructed in the milder climates.

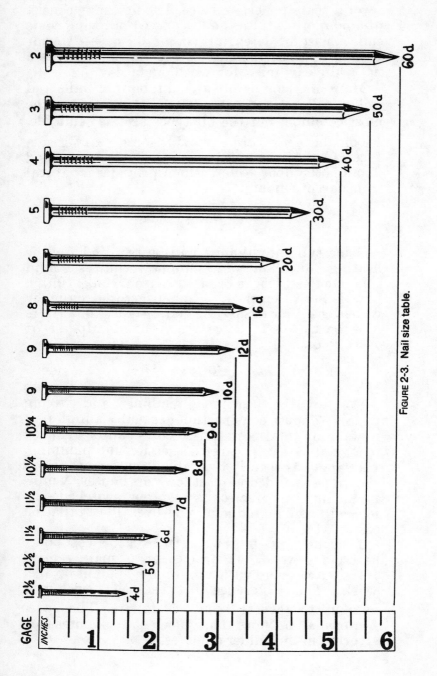

Figure 2-3. Nail size table.

Water supply and sewage disposal systems are often the most costly and difficult to install of all the utilities. When municipal or other systems are available for water and sewer service, there are no problems except the cost of installation. When good water can be obtained from a shallow well, a pump and pressure tank will provide a water supply at a low cost. However, in areas where a deep well is needed, costs may be too great. In such cases, one well might be used to supply several houses when they are ideally grouped. Wells should ordinarily be located a minimum of 50 feet from a septic tank and 100 feet from an absorption field.

Disposal of sewage in areas where public systems are available presents no problem. Where public systems are not available, the use of a septic tank and absorption system is required. The satisfactory performance of such a system depends on drainage, soil types, and other factors. It is sometimes necessary, when costs are critical, to provide for future installation of a disposal system. Roughed-in connections for plumbing facilities can and should be made in the house during its construction when sewer connections are not immediately available or costs are too great. Long-term costs of sewer connection will be more reasonable if provisions are made during construction.

Water supply and sewage disposal systems are specialized phases of house construction when these facilities are not available from a municipal or central source. Advice and guidance of a local health officer and engineer from your county office should be requested.

FOUNDATION SYSTEMS

One of the first essentials in house construction is to select the most desirable property site for its location. A lot in a smaller city or community presents few problems. The front set-back of the house and side-yard distances are either controlled by local regulations or should be governed by other houses in the neighborhood. However, if the site is in a rural or outlying area, care should be taken in staking out the house location.

Good drainage is essential. Be certain that natural drainage is away from the house or that such drainage can easily be assured by modification of ground slope. Low areas should be avoided. Soil conditions should be favorable for excavation for the treated posts or masonry piers of the foundation. Large rocks or other obstructions may require changes in the type of footings or *foundation*.

After the site for the house has been selected, all plant growth and sod should be removed. The area can then be raked and leveled slightly for staking and location of the supporting posts or piers.

The foundation plan in the working drawings for the house shows all the measurements necessary for construction. The first step in locating the house is to establish a baseline along one side with heavy cord and solidly driven stakes located well outside the end building lines (stakes 1 and 2, fig. 2-4). This baseline should be at the outer faces of the posts, piers, or foundation walls. When a post foundation is used with an overhang, the post faces will be 13½ inches in from the building line when a 12-inch overhang is used (fig. 2-5). When masonry piers or wood posts are located at the edge of the foundation, the outer faces are the same as the building line. These details are normally included in the working drawings. A second set of stakes (3 and 4) should now be established parallel to stakes 1 and 2 at the opposite side according to the measurements shown in the foundation plan of the working drawing. When measuring across, be sure that the tape is at right angles to the first baseline. Just as the 1-2 baseline does, this line will locate the outer edge of posts or piers. A third set of stakes, 5 and 6, should then be established at one end of the building line (fig. 2-4).

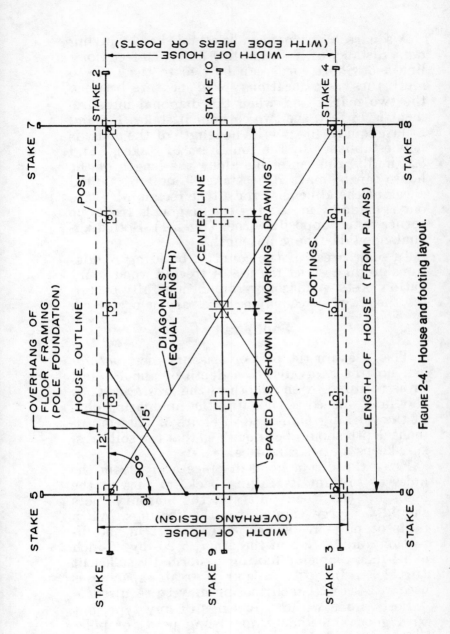

Figure 2-4. House and footing layout.

A square 90° corner can be established by laying out a distance of 12 feet at line 1–2 and 9 feet along line 5–6. Short cords can be tied to the lines to mark these two locations. Now measure between the two marks and when the diagonal measurement is 15 feet, the two corners at stakes 1–5 and 3–6 are square (fig. 2-4). The length of the house is now established by the fourth set of stakes, 7 and 8. Finally, the centerline along the length of the house can be marked by stakes 9 and 10. A final check of the alinement for a true rectangular layout is made by measuring the diagonals from one corner to the opposite corner (fig. 2-4). Both diagonals should be the same length.

In some areas of the country, building regulations might restrict the use of treated wood foundation posts. A masonry foundation fully enclosing the crawl space may be necessary or preferred.

Footings

The holes for the post or masonry pier *footings* can now be excavated to a depth of about 4 feet or as required by the depth of the *frostline*. They should be spaced as shown in the foundation plan of the working plans and in figure 5. The embedment depth should be enough so that the soil pressure keeps the posts in place.

Place the dirt a good distance away from the holes to prevent its falling back in. Size of the holes for the wood post and the masonry piers should be large enough for the footings. When posts or piers are spaced 8 feet apart in one direction and 12 feet in the other, a 20- by 20-inch or 24-inch-diameter footing is normally sufficient (fig. 2-5). In softer soils or if greater spacing is used, a 24- by 24-inch footing may be required.

Posts alone without footings of any type, but with good embedment, are being used for pole-type buildings. However, because the area of the bottom end of the pole against the soil determines its load capacity, this method is not normally recommended in the construction of a house where uneven settling could cause problems. A small amount of settling in a pole warehouse or barn would not be serious. Where soil capacities are very high and posts are spaced closely, it is likely that a footing support under the end of the post would not be required. However, because a good, stable foundation is important in any type of house, the use of adequate footings of some type must be considered.

After the holes have been dug to the recommended depth and cleared of loose dirt, an 8-inch-thick or thicker concrete footing should be poured (fig. 2-5). If premixed *concrete* is not available, it can be mixed by hand or by a small on-the-job mixer. Tops of the footings should be leveled by measuring down a constant distance from the level line. A 5- or 6-bag mix (premixed concrete) or a 1 to 2½ to 3½ (cement : sand : gravel) job-mixed concrete should be satisfactory. The footings for the post and pier foundations should be located as shown on the working drawings.

Post Foundations—With Side Overhang

Treatment of the foundation posts should conform to Federal Specification TT-W-571. Penetration of the preservative for foundation posts should be equal to one-half the radius and not less than 90 percent of the sapwood thickness. The selection of posts should be governed also by final finish and appearance. When cleanliness, freedom from odor, or paintability is essential, waterborne preservative-treated posts should be used. The important principle is not to use untreated posts in contact with the soil.

Treated posts having a top diameter as shown on the plans should be selected for each of the footing locations. The length can be determined by setting the layout strings to the level of the top of the beams and posts. Select the corner with the highest ground elevation and move the string on this stake to about 18 to 20 inches above the ground level at this point. The minimum clearance

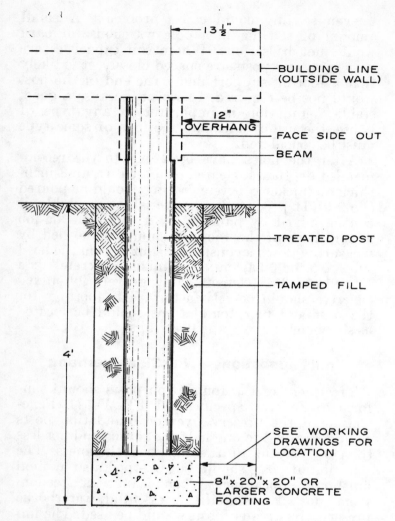

FIGURE 2-5. Post embedment and footing alinement (overhang design).

under joists or beams should be 12 inches. However, 18 to 24 inches or more is preferred when accessibility is desired. Then with the aid of a lightweight string or line level (fig. 2-6), adjust the cord on the other stakes so that the layout strings around the edge of the building and down the center are all truly level and horizontal. To

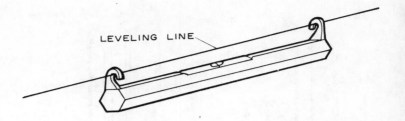

FIGURE 2-6. Line level. Locate line level midway between building corners when leveling.

insure accuracy, the line must be tight with no sag and the level located at the center. If available a surveyor's level will serve even better. Thus, the distance from the string to the top of concrete footings will now be the length of the posts needed at each location. A manometer-type level can also be used in establishing a constant elevation for the posts. This type of level consists of a long, clear plastic tube partly filled with water or other liquid. The water level at each end establishes the correct elevation.

If pressure-treated posts are not available in the lengths just determined, use poles more than twice as long as required. Saw them in half and use *with the treated end down.* Now with a saw and a hand ax or drawknife, slightly *notch* and face one side for a distance equal to the depth of the beams (fig. 2-7A). The four largest diameter posts should be used for the corners and notched on two adjacent sides (fig. 2-7B). Facing should be about 1½ inches wide, except for corners or when beam joints might be made where 2½ inches is preferable.

Treated 6- by 6-inch or 8- by 8-inch posts can be used in place of the round posts when available. Although they may cost somewhat more and do *not* have the resistance of treated round posts, square posts will reduce on-site labor time.

Locate the notched and faced posts on the concrete footings using the cord on the stakes as a guide for the faced sides. Place and tamp 8 to 10

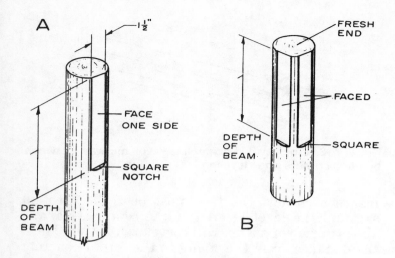

FIGURE 2-7. Facing posts: A, side or intermediate post; B, corner post (use largest).

inches of dirt around them initially to hold them in place. Posts should be vertical and the faced side alined with the cord from the stakes along each side, the center, and the ends of the house outline. When posts are alined, fill in the remaining dirt. Fill and tamp no more than 6 inches in the hole at one time to insure good, solid embedment.

Select *beams* the size and length shown in the foundation plan of the working drawings. Moisture content should normally not exceed 19 percent. These beams are usually 2 by 10 or 2 by 12 inches in size and the lengths conform to the spacing of the posts. For example, posts spaced 8 feet apart will require 8- and 16-foot-long beams. The outside beam can now be nailed in place. Starting along one side at the corner, even with the leveling string, nail one beam to the corner post and each crossing post (fig. 2-8*A*). Initially, use only one twentypenny nail at the top of each beam, and don't drive it in fully. (The center should be left free to allow for a carriage bolt.) Side beam ends should project beyond the post about 1½ inches or the thickness of the end header (fig. 2-8*B*). When all outside beams and those along the center row of posts are erected, all final leveling adjustments should be made. In addition to the leveling cord from the layout stakes, use a carpenter's level and a straightedge to insure that each beam is level, horizontal, and in line with the cord. Final nailing can now be done on the first set of beams. Posts extending over the tops of the beams can now be trimmed flush.

The second set of beams on the opposite sides of the posts should now be installed. Because round posts vary in diameter, the facing on the second side has been delayed until this time. Use a strong cord or string and stretch along the length of the side of the foundation on the inside of the posts and parallel to the outside beams (fig. 2-9A). This will establish the amount of notching and facing to be done for each post. Use a saw to provide a square notch to support the second beam. All posts can be thus faced and the second beams nailed in place level with, and in the same way as, the outside beams. This facing is usually unnecessary when square posts are used (fig. 2-9B). However, additional bolts are required when the beam does not bear on a notch. Joints of the *headers* should be made over the center of the posts and staggered. For example, if an 8- and a 16-foot beam are used on the outside of the posts, stagger the joints by first using a 16-foot, then an 8-foot length, on the inside. Only one *joint* should be made at each support.

Drill ½-inch holes through the double beams and posts at the midheight of the beam, and install ½- by 8-, 10-, or 12-inch galvanized carriage bolts with the head on the outside and a large washer under the nut on the inside. Use two bolts for square posts without a notch (fig. 2-9B). At splices, use two bolts and stagger (fig. 2-10A). When available at the correct moisture content, single nominal 4-inch-thick beams might be used to replace the two 2-inch members (fig. 2-10B).

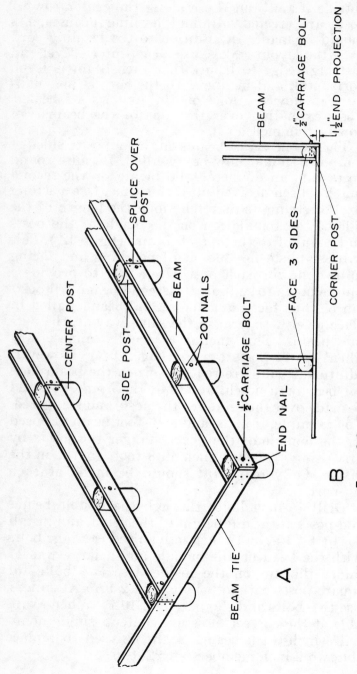

FIGURE 2-8. Beam installation: A, overall view; B, plan view.

All poles and beams are now installed and the final earth tamping can be done around the poles where required. A final raking and leveling is now in order to insure a good base for the soil cover if required. There is now a solid level framework upon which to erect the floor joists.

Edge Piers—Masonry and Posts

When masonry piers or wood posts are used along the edge of the building line instead of for overhang floor framing as previously described, the 8-inch poured footings are usually the same size as shown in fig. 2-10A. However, check the working drawings for the exact size. For masonry piers, the distance to the bottom of the footings should be governed by the depth of frost penetration. This may vary from 4 feet in the Northern States to less than 1 foot in the Southern areas. The wood posts normally require a 3- to 4-foot-deep hole. The masonry piers or posts should be alined so that the outside edges are flush with the outside of the building line (fig. 2-10A, B). The foundation plan in the working drawings covers these details further.

Concrete block, brick, or other masonry, or poured concrete piers can now be constructed over the footings. Concrete block piers should be 8 by 16 inches in size, brick or other masonry 12 by 12 inches, and poured concrete 10 by 10 inches. The tops of the piers should all be level and about 12 to 16 inches above the highest corner of the building area. Use any of the previously described leveling methods. Use a 22-gage by 2-inch-wide galvanized perforated or plain anchor strap for nailing into the beams. It should extend through at least two courses, filling the core when hollow masonry is used (fig. 2-10,A). A prepared mortar mix with 3 or 3½ parts sand and ¼ part cement to each part of mortar or other approved mixes should be used in laying up the masonry units. The wood post installation details are shown in figure 11B. Anchor straps are nailed to each post and beam with twelvepenny galvanized nails.

Beams consisting of doubled 2 by 10 or 2 by 12 members (check the foundation plan of the working drawings) can now be assembled. Place them on the posts or piers and make the splices at this location. Make only one splice at each pier. Nominal 2-inch members can be nailed together with tenpenny nails spaced 16 inches apart in two rows. Fabrication details at the corner and intersection with the center beam and fastening of the beam tie (*stringer*) are shown in figure 2-11.

Ledgers, used to support the floor joists, should be nailed to the inside of the nailed beams. The sizes are 2 by 2, 2 by 3, or 2 by 4 as indicated in the foundation plan of the working drawings. Ledgers should be spaced so that the top of the joists will be flush with the top of the edge and center beams when bearing on the ledgers (fig. 2-10A, B). Use sixteenpenny nails spaced 8 inches apart in a staggered row to fasten the ledger to the beam.

The foundation and beams are now in place ready for the assembly of the floor system.

An alternate method of providing footings for the treated wood foundation posts involves the use of temporary braces to position the posts while the concrete is poured (fig. 2-12). After the holes are dug, posts of the proper length are positioned and temporary braces nailed to them (fig. 2-12A). A minimum of 8 inches should be allowed for footing depth (fig. 2-12B). When posts are alined and set to the proper elevation, concrete is poured around them. After the concrete has set, holes are filled (*backfilled*) and earth tamped firmly around the posts. Construction of the floor framing can now begin.

This system of setting posts and pouring the concrete footings can also be used when a beam is located on each side of the posts (fig. 2-8). Posts are placed in the holes, the beams nailed in their proper position, and the beams alined and blocked to the correct elevation. After the concrete has set

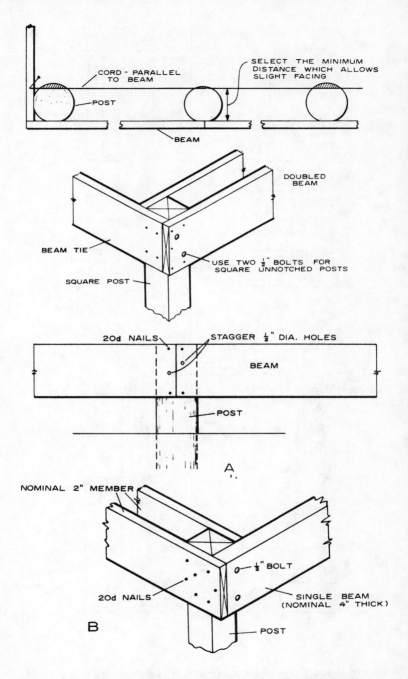

FIGURE 2-9. Facing and fastening round and square posts.

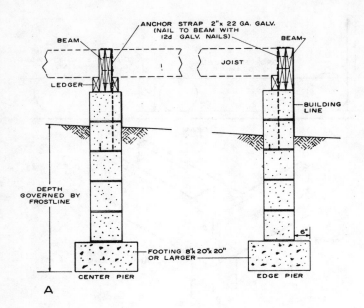

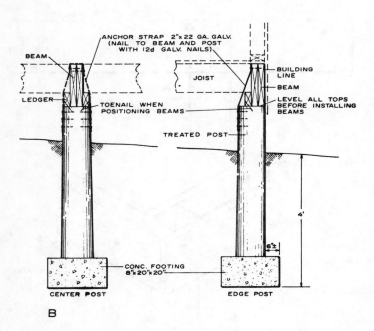

FIGURE 2-10. Edge posts and masonry piers. A, Masonry edge piers; B, edge post foundation.

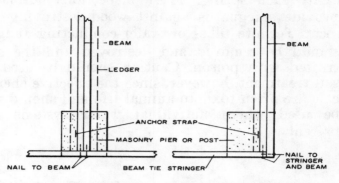

FIGURE 2-11. Corner and edge framing of beam.

and the fill has been tamped in place, the beams can be bolted to the posts.

Termite Protection

In areas of the South and in many of the Central and Coastal States, *termite* protection must be considered in construction of crawl-space houses. Pressure-treated lumber or poles are not affected by termites, but these insects build passages to and can damage untreated wood. Perhaps the most common and effective present-day method of protection is by the use of soil poisons. Spraying soil with solutions of approved chemicals such as aldrin, chlordane, dieldrin, and heptachlor using recommended methods will proved protection for 10 or more years.

A physical method of preventing entry of termites to untreated wood is by the use of *termite shields*. These are made of galvanized iron, aluminum, copper, or other metal. They are located over continuous walls (fig. 2-13A) or over or around treated wood foundation posts (fig. 2-13 B and C). They are not effective if bent or punctured during or after construction.

Crawl spaces should have sufficient room so that an examination of poles and piers can be made easily each spring. These inspections normally provide safeguards against wood-destroying insects. Termite tubes or water-conducting fungus should be removed and destroyed and the soil treated with poison. Caution should be used in soil treatment, however, since the effective chemicals are often toxic to animal life and should not be used where individual water systems are present.

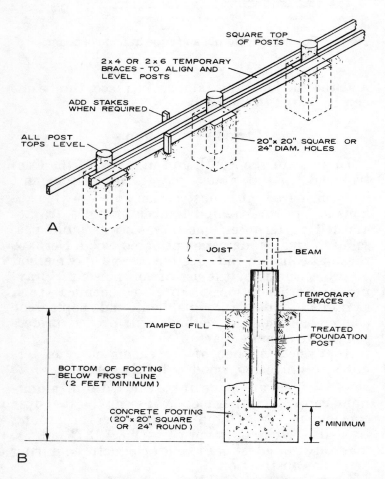

FIGURE 2-12. Alternate method of setting edge foundation posts. A, Temporary bracing; B, footing position.

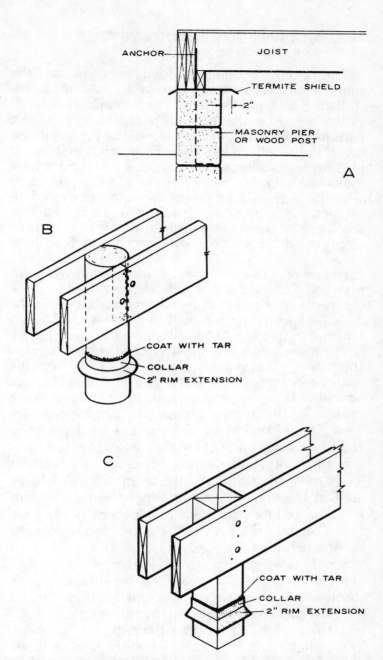

FIGURE 2-13. Termite shields: A, on top of masonry or wood post; B, round posts; C, square posts.

JOISTS FOR 3 FLOOR SYSTEMS

The beams, or beams and ledger strips, and the foundations are now in place and joists can be installed. Size and lengths of the floor joists, as well as the species and spacing, are shown in the floor framing layout in the working plans for the house being built. The joists may vary from nominal 2 by 8 inches in size to 2 by 10 inches or larger where spans are long. Moisture content of floor joists and other floor framing members should not exceed 19 percent when possible. Spacing of joists is normally 16 or 24 inches on center so that 8-foot lengths of plywood for subfloor will span six or four joist spaces.

In low-cost houses, savings can be made by using plywood for subfloor which also serves as a base for resilient tile or other covering. This can be done by specifying tongued-and-grooved edges in a plywood grade of C-C plugged Exterior Douglar-fir, southern pine, or similar species. Regular Interior Underlayment grade with exterior glue is also considered satisfactory. The *matched* edges provide a tight lengthwise joint, and end joints are made over the joists. If tongued-and-grooved plywood is not available, use square edged plywood and block between joists with 2 by 4's for edge nailing. Plywood subfloor also serves as a tie between joists over the center beam. Insulation should be used in the floor in some manner to provide comfort and reduce heat loss. It is generally used between or over the joists. These details are covered in the working drawings.

Single-floor systems can also include the use of nominal 1- by 4-inch matched finish flooring in species such as southern pine and Douglas-fir and the lower grades of oak, birch, and maple in $25/32$-inch thickness. To prevent air and dust infiltration, joists should first be covered with 15-pound asphalt felt or similar materials. The flooring is then applied over the floor joists and the flood insulation added when the house is enclosed. When this single-floor system is used, however, some surface protection from weather and mechanical damage is required. A full-width sheet of heavy plastic or similar covering can be used, and the walls erected directly over the film. When most of the exterior and interior work is done, the covering can be removed and the floor sanded and finished.

Post Foundation—With Side Overhang

The joists for a low-cost house are usually the third grade of such species as southern pine or Douglas-fir and are often 2 by 8 inches in size for spans of approximately 12 feet. If an overhang of about 12 inches is used for 12-foot lengths, the joist spacing normally can be 24 inches. Sizes, spacing, and other details are shown in the plans for each individual house.

The joists can now be cut to length, using a *butt joint* over the center beam. Thus, for a 24-foot-wide house, each pair of joists should be cut to a 12-foot length, less the thickness of the end header joist which is usually 1½ inches. The edge or stringer joists should be positioned on the beams with several other joists and the premarked headers nailed to them with one sixteenpenny nail (or just enough to keep them in position). The frame, including the edge (stringer) joists and the header joists, is now the exact outline of the house. Square up this framework by using the equal diagonal method (fig. 2-4). The overhang beyond the beams should be the same at each side of the house. Now, with eightpenny nails, *toenail* the joists to each beam they cross and the stringer joists to the beam beneath (fig. 2-14) to hold the framework exactly square. Add the remaining joists and nail the headers into the ends with three sixteenpenny nails. Toenail the remaining joists to the headers with eightpenny nails. When the center of a parallel partition wall is more than 4 inches from the center of the joists, add solid blocking between the joists. The blocking should be the same size as the

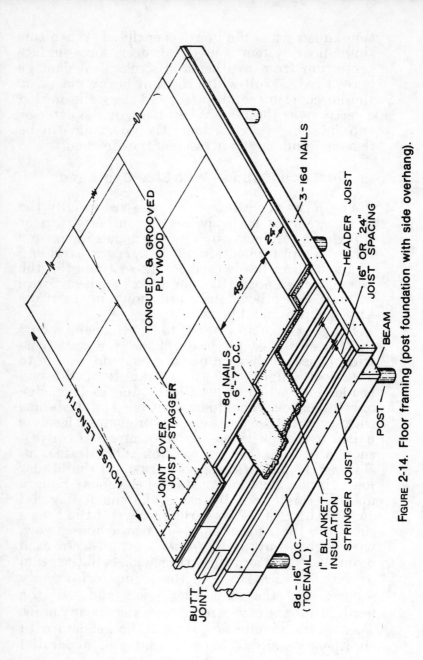

FIGURE 2-14. Floor framing (post foundation with side overhang).

joists and spaced not more than 3 feet apart. Toenail blocking to the joists with two tenpenny nails at each side.

In moderate climates, 1-inch blanket insulation may be sufficient to insulate the floor of crawl-space houses. It is usually placed between the joists in the same way that thicker insulation is normally installed. Another method consists of rolling 24-inch-wide 1-inch insulation, across the joists, nailing or stapling it where necessary to keep it stretched with tight edge joints (fig. 2-14). Insulation of this type should have strong damage-resistant covers. Tenpenny ring-shank nails should be used to fasten the plywood to the joists rather than eightpenny common normally used. This will minimize nail movement or "nail pops' which could occur during moisture changes. The vapor barrier of the insulation should be on the upper side toward the subfloor. Two-inch and thicker blanket or batt insulation is placed between the joists and should be applied any time after the floor is in place, preferably when the house is near completion.

When the house is 20, 24, 28, or 32 feet wide, the first row of tongued-and-grooved plywood sheets should be 24 inches wide, so that the butt-joints of the joists at the center beam are reinforced with a full 48-inch-wide piece (fig. 2-15). This plywood is usually 5/8- or 3/4-inch thick when it serves both as subfloor and underlayment. Rip 4-foot-wide pieces in half and save the other halves for the opposite side. Place the square, sawed edges flush with the header and nail the plywood to each crossing joist and header with eightpenny common nails spaced 6 to 7 inches apart at edges and at intermediate joists (fig. 2-14). Joints in the next full 4-foot widths of plywood should be broken by starting at one end with a 4-foot-long piece. End joints will thus be staggered 48 inches. End joints should always be staggered at least one joist space, 16 or 24 inches. Be sure to draw up the tongued-and-grooved edges tightly. A chalked snap-string

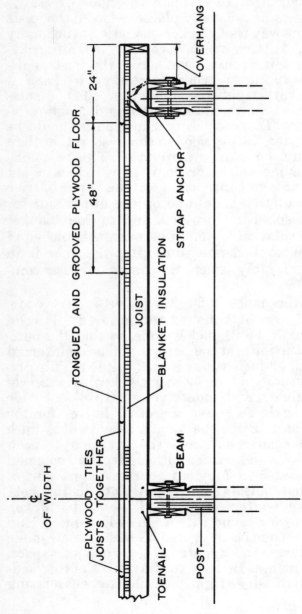

FIGURE 2-15. Floor framing details (post foundation).

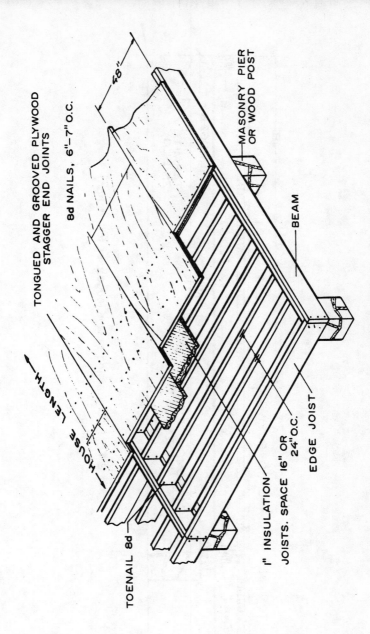

FIGURE 2-16. Floor framing for edge foundation (masonry piers or wood posts).

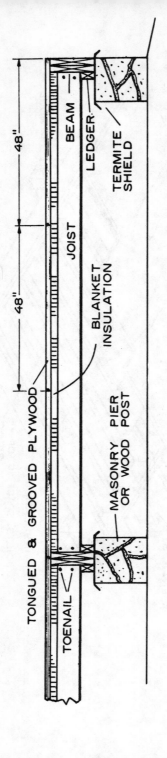

FIGURE 2-17. Floor framing details (edge pier foundation).

should be used to mark the position of the joists for nailing.

Edge Foundation-Masonry Pier or Wood Post

When edge piers or wood posts are used with edge support beams, the joists can be cut to length to fit snugly between the center and outside beams so that they rest on the ledger strips. Sizes of headers, joists, and other details are shown on the working drawings for each individual house. Toe-nail each end to the beams with two eightpenny nails on each side (fig. 2-16). In applying the plywood subfloor to the floor framing, start with full 4-foot-wide sheets rather than the 2-foot-wide pieces used for the side overhang framing (fig. 2-17). The nail-laminated center beam provides sufficient reinforcing between the ends of joists. Apply the insulation and nail the plywood the same way as outlined in the previous section.

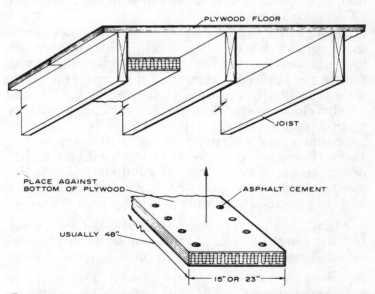

FIGURE 2-18. Application of friction-fit batt insulation between floor joists.

Insulation Between Joists

Thicker floor insulation than the 1-inch blanket is usually required for most houses. This will be indicated in the floor framing details of the working drawings or in the specifications. This type is normally used between the joists. Thus, the subfloor is nailed directly to the joists and the insulation placed between the joists after the subfloor is in place. Friction-fit (or similar insulating batts) 15 inches wide should be used for joists spaced 16 inches on center. Use 23-inch-wide batts for joists spaced 24 inches on center.

Friction-fit batts need little support to keep them in place. Small "dabs" of asphalt roof cement on the upper surfaces when installing against the bottom of the plywood will keep them in place (fig. 2-18). Standard batts can also be placed in this manner, but somewhat closer spacing of the cement might be required in addition to stapling along the edges. The use of a vapor barrier under the subfloor is important and is described in the section on "Thermal Insulation."

When other types of subfloor are specified, such as diagonal boards, some kind of overlay or finish is usually required. If tongued-and-grooved flooring is applied directly to and across the joists, a tie is normally required at the center butt joints of the floor joists. This is accomplished with a metal strap across the top of the joists or 1- by 4- by 20-inch wood strips (*scabs*) nailed across the faces of each set of joists at the joint with six eightpenny nails. When plywood subfloor is used, the sheets are centered over the center beam and joist ends to provide this tie for overhang floor framing.

Finally, if the plywood is likely to be exposed for any length of time before enclosing the house, a brush coat (or squeegee application) of water-repellent preservative should be used. This will not only repel moisture but will prevent or minimize any surface degradation.

FRAMED WALL SYSTEMS

Exterior sidewalls, and in some designs an interior wall, normally support most of the roof-loads as well as serving as a framework for attaching interior and exterior coverings. When roof trusses spanning the entire width of the house are used, the exterior sidewalls carry both the roof and ceiling loads. Interior partitions then serve mainly as room dividers. When ceiling joists are used, interior partitions usually sustain some of the ceiling loads.

The exterior walls of a wood-frame house normally consist of *studs*, interior and exterior coverings, windows and doors, and insulation. Moisture content of framing members usually should not exceed 19 percent.

The framework for a conventional wall consists of nominal 2- by 4-inch members used as top and bottom *plates*, as studs, and as partial (cripple) studs around openings. Studs are generally cut to lengths for 8-foot walls when subfloor and finish floors are used. This length depends on the thickness and number of wall plates (normally single bottom and double top plates). Studs can often be obtained from lumber dealers in a precut length. Double headers over doors and windows are generally larger than 2 by 4's when the width of the opening is greater than $2\frac{1}{2}$ feet. Two 2- by 6-inch members are used for spans up to $4\frac{1}{2}$ feet and two 2- by 8-inch members for openings from $4\frac{1}{2}$ to about $6\frac{1}{2}$ feet. Headers are normally cut 3 inches (two $1\frac{1}{2}$-inch stud thicknesses) longer than the rough opening width unless the edge of the opening is near a regular spaced stud.

Framing of Sidewalls

The exterior framed walls, when erected, should be flush with the outer edges of the plywood subfloor and floor framing. Thus, the floor can be used both as a layout area and for horizontal assembly of the wall framing. When completed, the

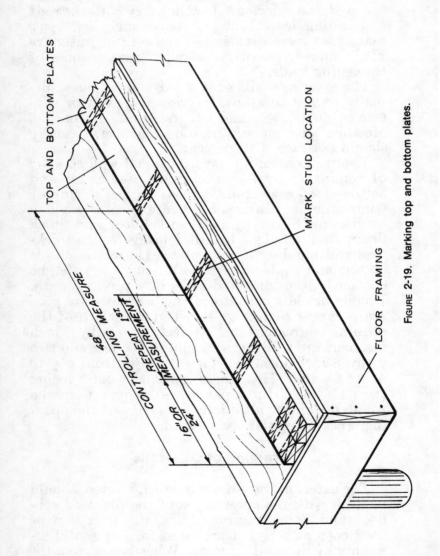

FIGURE 2-19. Marking top and bottom plates.

entire wall can be raised in place in "tilt-up" fashion, the plates nailed to the floor system, and the wall *plumbed* and braced.

The two exterior sidewalls of the house can be framed first and the exterior end walls later. Cut two sets of plates for the entire length of the house, using 8-, 12-, or 16-foot lengths, staggering the joints. Joints should be made at the centerline of a stud (*on center*). Starting at one end, mark each 16 or 24 inches, depending on the spacing of the studs, and also mark the centerlines for windows, doors, and partitions. These measurements are given on the working drawings. These are centerline (*o.c.*) markings, except for the ends. With a small square, mark the location of each stud with a line about ¾ inch on each side of the centerline mark (fig. 2-19).

Studs can now be cut to the correct length. When a low-slope roof with wood decking (which also serves as a ceiling finish) is used, the stud length for an 8-foot wall height with single plywood flooring should be 95½ inches, less the thickness of three plates. Thus for plates 1½ inches thick, this will mean a stud length of 91 inches. When ceiling joists or trusses are used with the single plywood floor, this length can be about 92⅛ inches. These measurements are primarily to provide for the vertical use of 8-foot lengths of dry-wall sheet materials for the walls with the ceiling finish in place. Cornerposts can be made up beforehand by nailing two short 2- by 4-inch blocks between two studs (fig. 2-20). Use two twelvepenny nails at each side of each block.

Begin fabrication of the wall by fastening the bottom plate and the first top plate to the ends of each cornerpost and stud (with two sixteenpenny nails) into each member. As studs are nailed in place, provisions should be made for framing the openings for windows and doors (fig. 2-20). Studs should be located to form the rough openings, the sizes of which vary with the types of windows selected. The rough openings are framed

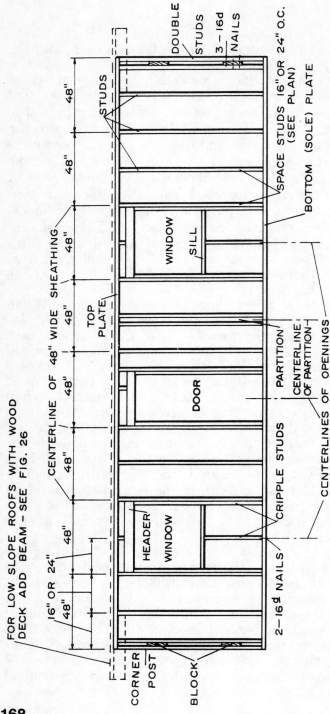

Figure 2-20. Framing layout of typical wall.

by studs which support the window and door headers or *lintels*. A full-length stud should be located at each side of these framing studs (fig. 2-21).

The following allowances are usually made for rough opening widths and heights for doors and windows. Half of the given width should be marked on each side of the centerline of the opening, which has previously been marked on top and bottom plates.

A. *Double-hung window* (*single unit*)
  Rough opening width = glass width *plus* 6 inches
  Rough opening height = total glass height *plus* 10 inches
B. *Casement window* (*two sash*)
  Rough opening width = total glass width *plus* 11¼ inches
  Rough opening height = total glass height *plus* 6⅜ inches

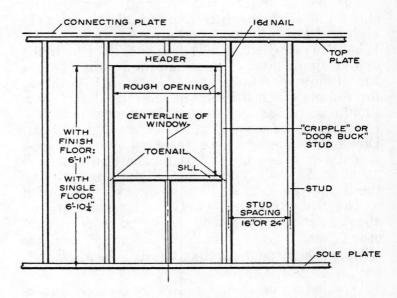

FIGURE 2-21. Framing at window opening and height of window and door headers.

C. *Exterior doors*
    Rough opening width = width of door *plus* 2½ inches
    Rough opening height = height of door *plus* 3 inches

Clearances, or rough-opening sizes, for typical double-hung windows, for example, are shown in table 2.

TABLE 2.—*Frame opening sizes for double hung windows*

Window glass size (each sash)		Rough frame opening size	
Width *Inches*	Height *Inches*	Width *Inches*	Height *Inches*
24	16	30	42
28	20	34	50
32	24	38	58
36	24	42	58

The height of the window and door headers above the subfloor when doors are 6 feet 8 inches high and finish floor is used is shown in fig. 2-21. When only resilient title is used over flooring made up of a single layer of material, the framing height for windows and doors should be 6 feet 10¼ inches for 6-foot 8-inch doors. The sizes of the headers should be the same as those previously outlined. Framing is arranged as shown in figure 2-21. Doubled headers can be fastened in place with two sixteenpenny nails through the stud into each member. Cripple (door *buck*) studs supporting the header on each side of the opening are nailed to the full stud with twelvepenny nails spaced about 16 inches apart and staggered. The sill and other short (cripple) studs are toenailed in place with two eightpenny nails at each side when end-nailing is not possible.

Doubled studs should normally be used on exterior walls where intersecting interior partitions are located. This is often accomplished with spaced studs, (fig. 2-22*A*) and provides nailing surfaces for interior covering materials. Blocking with 2-by-4-inch members placed flatwise between studs spaced 4 to 6 inches apart in the exterior wall might also be used to fasten the first partition stud (fig. 2-22*B*). Blocks should be spaced about 32 inches apart. When a low-slope roof with gable overhang is used with wood decking, a beam extension is required at the top plates (figs. 2-20 and 2-26).

Erecting Sidewalls

When the sidewalls are completed, they can be raised in place. Nail several short 1- by 6-inch pieces to the outside of the beam to prevent the wall from sliding past the edge. The bottom plate is fastened to the floor framing with sixteenpenny nails spaced 16 inches apart and staggered when practical. The wall can now be plumbed and temporary bracing added to hold it in place in a true vertical position. Bracing may consist of 1- by 6-inch members nailed to one face of a stud and to a 2 by 4 block which has been nailed to the subfloor. Braces should be at about a 45° angle. If the wall framing is squared and braced, the panel siding or exterior covering can be fastened to the studs while the walls are still on the subfloor. In addition, window frames can be installed before erection of the wall. These processes are covered in following sections on "Exterior Wall Coverings" and "Exterior Frames."

End Walls—Moderate-Slope Roof

The exterior end walls for a gable-roofed house may be assembled on the floor in the same general manner as the sidewalls with a bottom plate and single top plate. However, the total length of the wall should be the exact distance between the inside of the exterior sidewalls already erected. Fur-

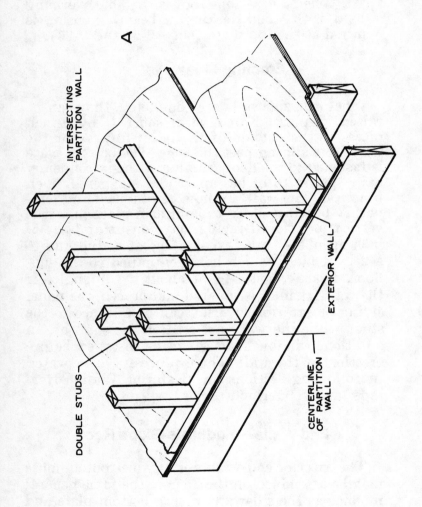

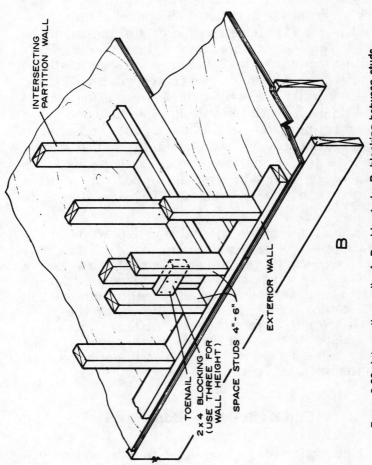

Figure 2-22. Intersecting walls. A, Double studs; B, blocking between studs.

thermore, only one end stud is used rather than the doubled cornerposts (fig. 2-23). Window and door openings are framed as outlined for the exterior sidewalls. When 48-inch-wide panel siding is used for the exterior, serving both as sheathing and siding, for example, the stud spacing should conform to the type of covering used. The center of the second stud in this wall should be 16 or 24 inches from the outside of the panel-siding material (fig. 2-24). This method of spacing should be used from each corner toward the center, and any adjustments required because of sheet-material width should be made at a center window or door.

End walls are erected in the same manner as the sidewalls with the bottom plate fastened to the floor framing. These walls also must be plumbed and braced. The end studs should be nailed at each side to the cornerposts with sixteenpenny nails spaced 16 inches apart. The upper top plate is added and extends across the sidewall plate (fig. 2-23).

The framing for the gable-end portion of the wall is often separately (fig. 2-23). Studs may be toenailed to the upper top wall plate, or an extra 2- by 6-inch bottom plate can be used, which provides a nailing surface for ceiling material in the rooms below. The top members of the gable wall are not plates; they are rafters which form the slope of the roof. Studs are notched to fit or may be used flatwise. Use the roof slope specified in the working drawings.

End Walls—Low-Slope Roof

End walls for a low-slope roof are normally constructed with *balloon framing*. In this design, the studs are full length from the bottom plate to the top or rafter plates which follow the roof slope (fig. 2-25). The stud spacing and framing for windows are the same as previously outlined. The

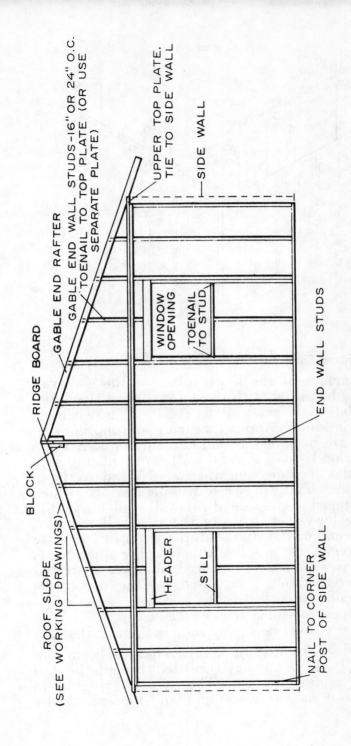

FIGURE 2-23. End wall framing for regular slope foof (for trusses or rafter-type).

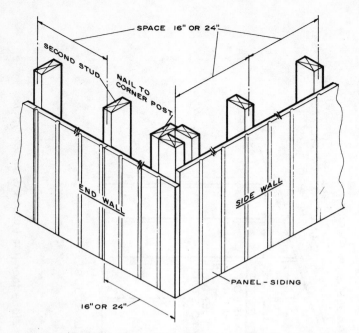

FIGURE 2-24. Exterior side and end wall intersection.

top surface of the upper plate of the end wall should be in line with the outer edge of the upper top plate of the sidewall (fig. 2-26). Thus, when the roof decking is applied, bearing and nailing surfaces are provided at end and sidewalls. A beam extension beyond the end wall is provided to support the wood decking when a *gable*-end overhang is desired. This can be a 4- by 6-inch member which is fastened to the second sidewall stud (fig. 2-26).

The lower top plate of the end wall is nailed to the end of the studs before the upper top plate is fastened in place. For attaching the upper plate, use sixteenpenny nails spaced 16 inches apart and staggered. Two nails are used over the cornerposts of the sidewalls (fig. 2-26).

To provide for a center *ridge* beam which supports the wood decking inside the house, the area should be framed (fig. 2-27). After the beam is in place, twelvepenny nails are used through the stud on each side of the beam. The size of the ridge beam is shown on the working plans for each house

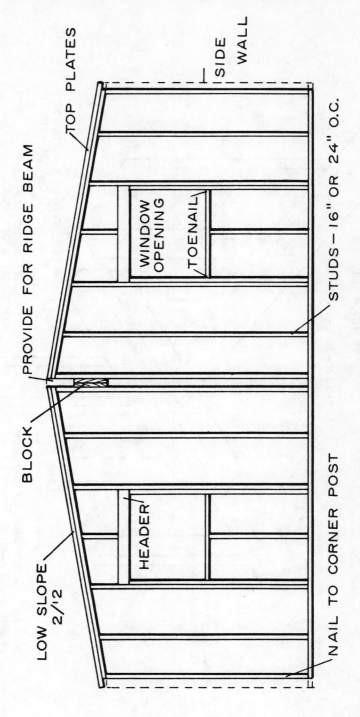

Figure 2-25. Framing for end wall (low-slope roof).

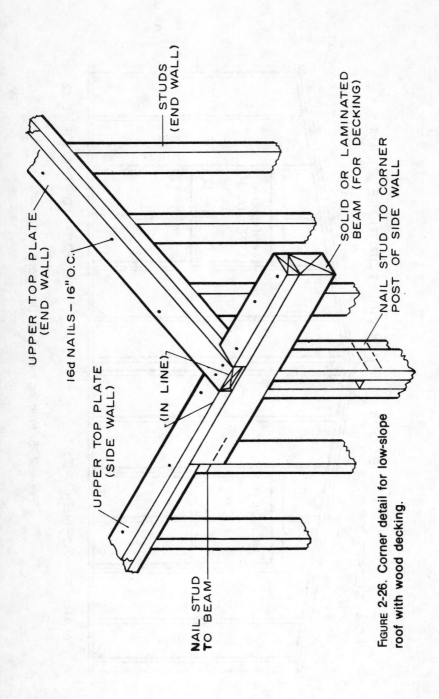

Figure 2-26. Corner detail for low-slope roof with wood decking.

which is constructed using this method. When decking is used for a gable overhang, the beam extends beyond the end walls.

Interior Walls

Interior walls in conventional construction (with ceiling joists and rafters) are erected in the same manner and at the same height as the outside walls. In general, assembly of interior stud walls is the same as outlined for exterior walls. The center load-bearing partition should be located so that ceiling joists require little or no wasteful cutting. Cross partitions are usually not load-bearing and can be spaced as required for room sizes. These spacings and other details are covered in the working drawing floor plan.

Studs should be spaced according to the type of interior covering material to be used. When studs are spaced 24 inches on center, the thickness of gypsum board, for example, must be ½ inch or greater. For 16-inch stud spacing, a thickness of ⅜ inch or greater can be used. Details of a typical intersection of interior walls are shown in fig. z-28. Load-bearing partitions should be constructed with nominal 2- by 4-inch studs, but 2- by 3-inch studs may be used for nonload-bearing walls. Doorway openings can also be framed with a single member on each side in nonload-bearing walls (fig. 2-28). Single top plates are commonly used on nonload-bearing interior partitions.

ROOF SYSTEMS

Roof trusses require no load-bearing interior partitions, so location of the walls and size and spacing of studs are determined by the house design and by the type of interior finish. The bottom chords of the trusses are often used to tie in with crossing partitions where required. Details of stud

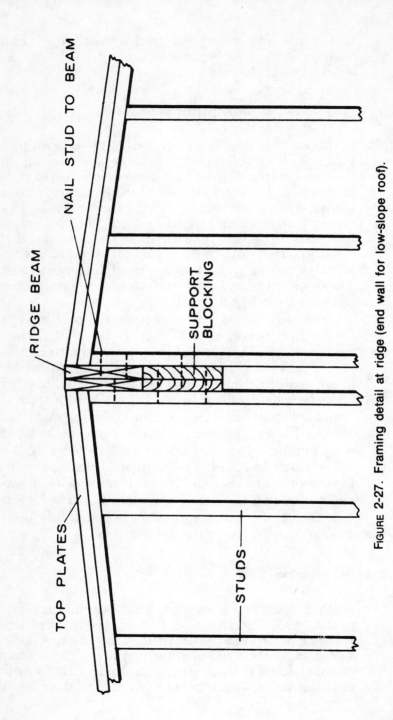

Figure 2-27. Framing detail at ridge (end wall for low-slope roof).

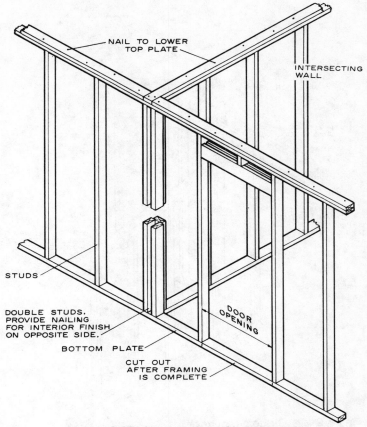

FIGURE 2-28. Intersection of interior walls.

location at the intersection of an interior partition with an exterior wall are shown in fig. 2-22 *A* and *B*.

When a low-slope roof is used with wood decking, a full-height wall or a ridge beam is required for support at the center (fig. 2-29). The ridge beam may span from an interior center partition to an outside wall, forming a clear open area beneath. Cross or intersecting walls are full height with a sloping top plate. In such designs, one method uses the following sequence: (a) Erect exterior walls, center wall, and ridge beam; (b) apply roof decking; and (c) install other partition walls.

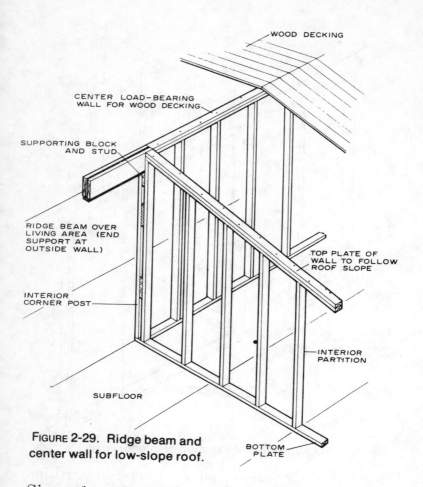

FIGURE 2-29. Ridge beam and center wall for low-slope roof.

Size and spacing of the studs in the cross walls are usually based on the thickness of the covering material, as no roof load is imposed on them. These spacings and sizes are part of the working drawings.

The upper top plates (connecting plates) are used to tie the wall framing together at corners, at intersections, and at crossing walls. The upper plate crosses and is nailed to the plate below (figs. 2-26 and 2-28). Two sixteenpenny nails are used at each intersection. The remainder of the upper top plate is nailed to the lower top plate with sixteenpenny nails spaced 16 inches apart in a staggered pattern.

The primary function of a roof is to provide protection to the house in all types of weather with a minimum of maintenance. A second consideration is appearance; a roof should add to the attractiveness of the home, as well as being practical. Happily, a roof with a wide overhang at the cornice and the gable ends not only enhances appearance, but provides protection to side and end walls. Thus, even in lower cost houses, when the style and design permit, wide overhangs are desirable. Though they add slightly to the initial cost, savings in future maintenance usually merit this type of roof extension. Wood members used for roof framing should normally not exceed 19 percent moisture content.

As briefly described in the section on "Major House Parts," the two types of roofs commonly used for houses are (a) the low-slope and (b) the *pitched* roof. The flat or low-slope roof combines ceiling and roof elements as one system, which allows them to serve as interior finish, or as a fastening surface for finish, and as an outer surface for application of the roofing. The structural elements are arranged in several ways by the use of ceiling beams or thick roof decking, which spans from the exterior walls to a ridge beam or center bearing partition. Roof slope is usually designated as some ratio of 12. For example, a "4 in 12" roof slope has a 4-foot vertical rise for each 12 feet of horizontal distance.

The pitched roof, usually in slopes of 4 in 12 and greater, has structural elements in the form of (a) *rafters* and joists or (b) trusses (trussed rafters). Both systems require some type of interior ceiling finish, as well as roof sheathing. With slopes of 8 in 12 and greater, it is possible to include several bedrooms on the second floor when provisions are made for floor loads, a stairway, and windows.

Low-Slope Ceiling Beam Roof

One of the framing systems for a low-slope roof consists of spaced rafters (beams or *girders*) which span from the exterior sidewalls to a ridge beam or a center load-bearing wall. The rafter can be doubled, spaced 4 feet apart, and exposed in the room below providing a pleasing beamed ceiling effect. *Dressed and matched* V-groove boards can be used for roof sheathing and exposed to the room below. When plywood or other unfinished sheathing is used, a ceiling tile or other prefinished wallboard can be fastened to the undersurface. Such materials also serve as insulation. Thus, a very attractive ceiling can be provided using a light color for the ceiling and a contrasting stain on the beams. This type of framing can be varied by spacing single rafters on 16- or 24-inch centers. Separate covering materials would normally be used for the roof sheathing and for the ceiling, with flexible insulation between.

The size and spacing details for ceiling beams are shown on the working drawings for each house design. For example, when beams are doubled, spaced 48 inches apart, and the distance from outer wall to interior wall is about 11½ feet, two 2- by 8-inch members are satisfactory for most of the construction species such as Douglas-fir, southern pine, and hemlock. Use of some wood species, such as the soft pines, will require two 2- by 10-inch members for 48-inch spacing. When a spacing of 32 inches is desirable for appearance, two 2- by 6-inch members of the second grade of Douglas-fir or southern pine are satisfactory. In some of the species, such as southern pine and Douglas-fir, a solid 4- by 6-inch member provides sufficient strength for 48-inch spacing over an 11½-foot span.

The details of fastening and anchoring these structural members to the wall elements can vary somewhat. Variations from the details included in this manual are shown on the working drawings for each individual house.

Construction.—We will assume that the details in the working plans specify doubled 2- by 8-inch ceiling beams spaced 48 inches on center. When the beams do not extend beyond the wall, a lookout member is required for the *cornice* overhang. The center wall or ridge beam is in place, so the roof slope, which may vary between 1½ in 12 to 2½ in 12, has been established. As previously outlined in the section on wall systems, the ceiling beams should normally be erected before cross walls are established. Thus, the exterior sidewalls and the load-bearing center wall, all well braced and plumbed, are all that is required to erect the ceiling beams.

There are several methods in which the beams are supported at the center bearing wall or ridge beam: (a) By a 2- by 3-inch block fastened to the stud wall; (b) by a metal joist hanger; and (c) by notching the beam ends and fastening to the stud. The first two may be used for either a stud wall or a ridge beam. Fastening at the outside wall is generally the same for all three methods.

The first or sample beam can now be cut to serve as a pattern in cutting the remainder of the members. Fig. 2-30*A* shows the location of the ceiling beam with respect to the exterior and load-bearing center walls. When beams themselves do not serve as roof extensions, they can be assembled by nailing a 2- by 6-inch *lookout* (roof extension) at the outer wall and a 2- by 4-inch spacer block at the interior center wall (fig. 2-30*B*). Use two twelvepenny nails on each side of the block and twelvepenny nails spaced 6 inches apart for the 2- by 6-inch lookout member. When a block nailed to the stud is used to support the interior beam end, the ends should be notched (fig. 2-30*C*).

Details at center wall or beam.—The first system of connecting the inside end of the ceiling beam is most adaptable to ridge-beam construc-

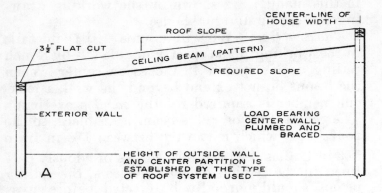

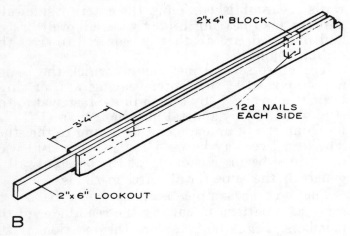

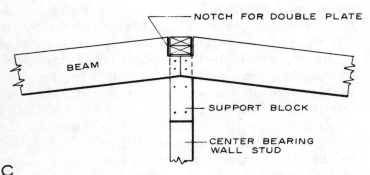

FIGURE 2-30. Ceiling beam location and support methods. A, Layout of typical beam; B, assembly of beam with roof extension (lookout); C, notch for stud support at center bearing wall.

tion. It consists of fastening a nominal 2- by 3-inch block to the beam with 4½-inch lag screws (fig. 2-31A). The 2 by 3 should be the same depth as the ceiling beam. The beam ends are then bolted to the 2- by 3-inch block with ⅜- by 5-inch carriage bolts.

Using joist hangers (fig. 2-31B) to fasten the inside end of the ceiling beams is another method most adaptable to a ridge-beam system. Hangers are fastened to the ridge-beam face, the ceiling beams dropped in place, and the hangers nailed to the beams. Eight- or tenpenny nails are commonly used for nailing. Hangers will be exposed, but can be painted to match the color of the beams.

A third method which can be used at a center bearing wall includes notching the ends of the ceiling beams (figs. 2-30C and 31C). When the beam is in place, it is face-nailed to the stud at each side with two twelvepenny nails. Short 2- by 4-inch support blocks are then nailed at each side of the stud with twelvepenny nails (fig. 2-31C).

When solid 4- by 6-inch or larger members are used as ceiling beams in place of the doubled members, the joist hanger is probably the most suitable method of supporting the inside ends of the beams. If the solid or laminated beams are to be stained, care should be taken to prevent hammer marks.

Nailing ceiling beams at exterior walls.—The ceiling beams are normally fastened to the top plate of the outside walls by nailing. In windy areas, some type of strapping or metal bracket is often desirable (fig. 2-41B). The ceiling beams are toenailed to the top of the outside wall with two eightpenny nails at each side and a tenpenny nail at the ends (fig. 2-32). To provide nailing for panel siding and interior finish, 2- by 4-inch nailing blocks are fastened between the ceiling beams (fig. 2-32). Toenail with eightpenny nails at each edge and face.

Roof sheathing.—Ceiling beams and roof extensions are now in place and ready for installation

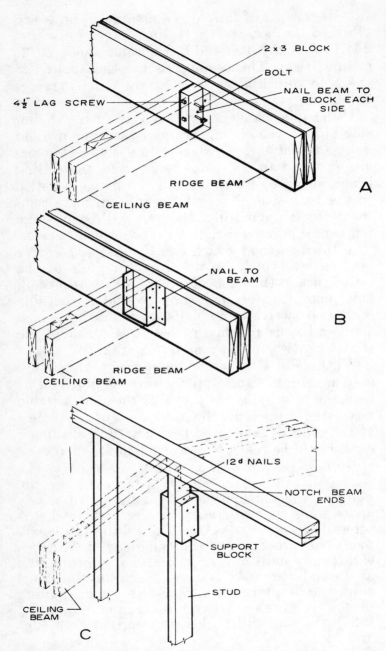

FIGURE 2-31. Beam connection to ridge-beam or load-bearing wall. A, Block support; B, joist hanger; C, stud support block.

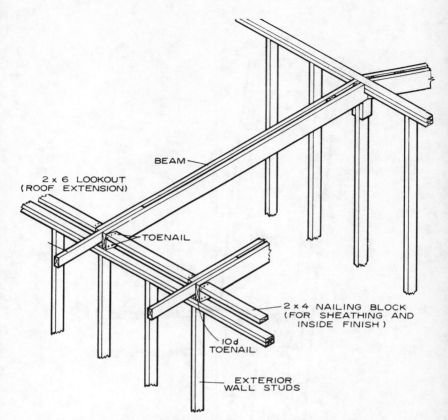

FIGURE 2-32. Fastening ceiling beams at exterior walls.

of the roof sheathing. Roof sheathing can consist of 1- by 6-inch tongued-and-grooved V-groove) lumber with $25/32$-inch fiberboard nailed over the top for insulation. Use two eightpenny nails for each board at each ceiling beam. The insulation fiberboard can be nailed in place with $1\frac{1}{4}$-inch roofing nails spaced 10 inches apart in rows 24 inches on center. Cross sections of the completed wall and roof framing are shown in figure 2-33 A and B. A nominal 1-inch member about $4\frac{3}{4}$ inches wide may be used to case the undersides of the beams.

A gable-end extension of 16 inches or less can be supported by extending the dressed and

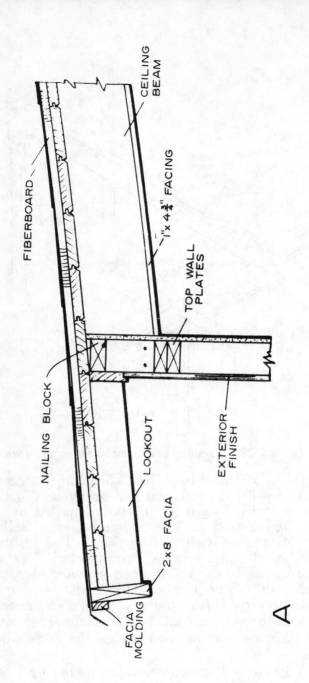

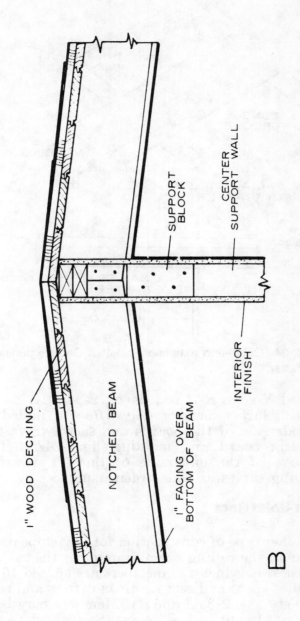

FIGURE 2-33. Cross sections of completed walls and roof framing. A, Section through exterior wall; B, section through center wall.

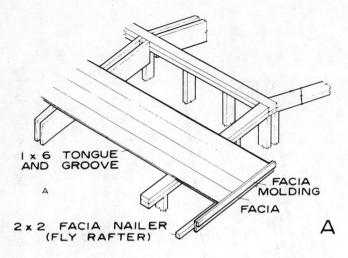

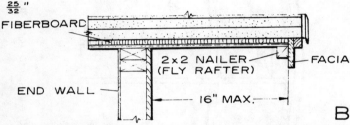

FIGURE 2-34. Gable-end extension detail. A, Gable extension; B, fly rafter.

matched V-edge roof boards (fig. 2-34*A*). A 2- by 2-inch or larger member (*fly rafter*) is nailed to the underside of the boards and serves to fasten the facia board and molding (fig. 2-34*B*). The V-groove of the underside of the 1 by 6 roof sheathing serves as a decorative surface.

Rafter-Joist Roof

Another type of construction for low-slope roofs similar to the ceiling beam framing is the rafter-joist roof, in which the members are spaced 16 or 24 inches apart and serve both as rafters and ceiling joists (fig. 2-35 *A* and *B*). Members may be 2 by 8 or 2 by 10 inches in size. Specific sizes are shown on the working drawings for each house plan. The space between joists is insulated, allowing space for a ventilating *airway*. Gypsum board or other types of interior finish can be nailed directly to the bottoms of the joists.

Rafter extensions can serve as nailing surfaces for the *soffit* of a closed cornice(fig.2-35*A*). When an open cornice is used, a nailing block is required over the wallplates and between rafters for the siding or *frieze* board.

The inside ends of the rafter-joists bear on an interior load-bearing wall (fig. 2-35*B*). Beams are toenailed to the plate with eightpenny nails on each side. A 1- by 4-inch wood or ⅜-inch plywood *scab* is used to connect opposite rafter-joists. This fastens the joists together and serves as a positive tie between the exterior sidewalls.

Low-Slope Wood-Deck Roof

A simple method of covering low-slope roofs is with wood decking. Decking should be strong enough to span from the interior center wall or beam to the exterior wall. Decking can also extend beyond the wall to form an overhang at the *eave* line (fig. 2-36*A*). This system requires dressed and matched nominal 2- by 6-inch southern pine or Douglas-fir decking or 3- by 6-inch solid or laminated decking (cedar or similar species) for spans of about 12 feet. The proper sizes are shown in the working drawings. While this system requires more material than the beam and sheathing system, the labor involved at the building site is usually much less.

When gable-end extension is desired, some type of support is required beyond the end wall line at plate and ridge. This is usually accomplished by the projection of a small beam at the top plate of each sidewall (figs. 2-20 and 26) and at the ridge. Often the extension of the double top plate of the sidewall is sufficient. Depending on the type and thickness, the decking must sometimes be in one full-length piece without joints unless there are intermediate supports in the form of an interior partition. When such an interior wall is present,

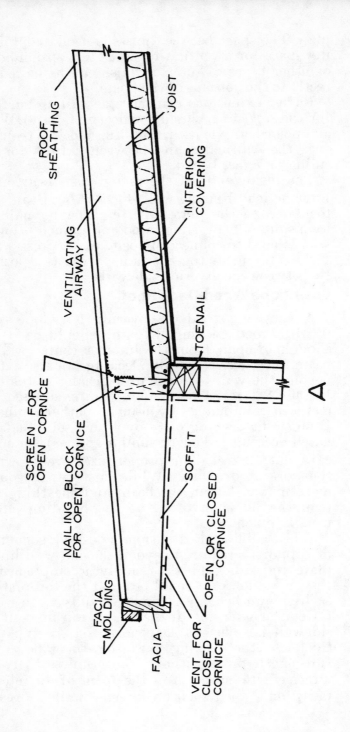

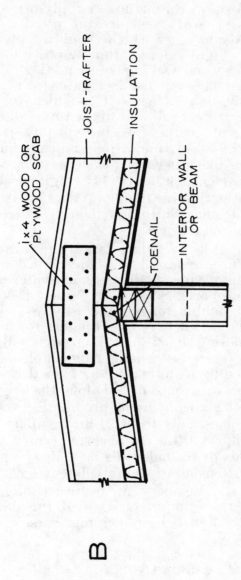

FIGURE 2-35. Rafter-joist construction. A, Detail at exterior wall; B, detail at interior wall.

a butt joint can be made over its center. The working drawings cover these various details.

Figure 2-36 shows the method of applying wood decking. This type of wood decking usually has a decorative V-edge face which should be placed down. Often a light-colored stain or other finish is applied to the wood decking before it is installed. Prefinished members can also be obtained. Each 2- by 6-inch decking member is face-nailed to the ridge beam or center wall and to the top plates of the exterior wall with two sixteenpenny ring-shank nails (fig. 2-36A). In addition, sixpenny finish nails should be toenailed along each joint on 2- to 3-foot centers (fig. 2-36B). A 40° angle or less should be used so that the nail point does not penetrate the underside. Nailheads should be driven flush with the surface.

When nominal 3- by 6-inch decking in solid or laminated form is required, it is face-nailed with two twentypenny ring-shank nails at center and outside wall supports. Solid 3- by 6-inch decking usually has a double tongue and groove and is provided with holes between the tongues for horizontal edge nailing (fig. 2-36C). This edge nailing is done with 7- or 8-inch-long ring-shank nails, often furnished by manufacturers of the decking. Laminated decking can be nailed along the lengthwise joints with sixpenny nails through the groove and tongue. Space nails about 24 inches apart.

Decking support at the load-bearing center wall and at the ends of the sidewalls may also be provided by an extension of the top plates (fig. 2-37). The sides of the members can be faced later, if desired, with the same material used for siding or with 1- by 6- and 1- by 8-inch members.

Insulation for Low-Slope Roofs

It is somewhat more difficult and costly to provide the low-slope roof with a great amount of in-

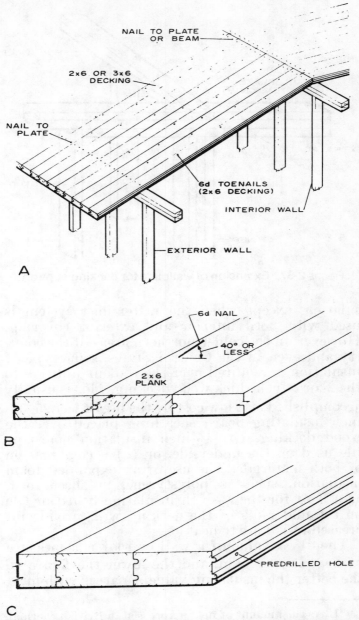

FIGURE 2-36. Wood deck construction. A, Installing wood decking; B, toenailing horizontal joint; C, edge nailing 3- by 6-inch solid decking.

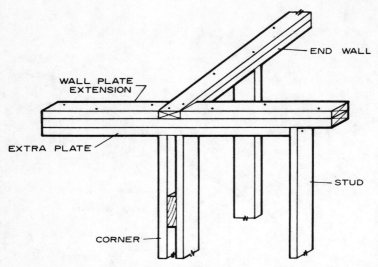

FIGURE 2-37. Extension of wallplate for decking support.

sulation, except when the rafter-joist system is used with both interior and exterior covering. However, in the colder northern areas of the country and even in the Central States, some type of insulation is required over the roof in addition to the 2- or 3-inch-thick wood decking. This is usually accomplished at a low cost by the use of (a) 25/32-inch insulating board sheathing placed over the wood decking, (b) ½-inch insulation board or tile used on the underside, or (c) a combination of both materials. The use of an expanded foam insulation, such as polystyrene, in sheet form as a base for the ½-inch insulation board or tile on the underside of the decking will provide increased resistance to heat loss.

The "U"[1] value of a wall or roof is a measure of heat loss or gain, and the lower the figure is, the better the insulating value. In areas with mild

---

[1] "U" is the amount of heat, expressed in British thermal units, transmitted in 1 hour through 1 square foot of surface per 1° F. difference in temperature between the inside and outside air.

winters, the use of the 2- by 6- or 3- by 6-inch decking alone should be sufficient. In moderate temperature areas, the use of the 3- by 6-inch cedar or similar decking alone, or the 2- by 6-inch decking with a cover of 25/32-inch insulating board sheathing, is probably sufficient. In the northern tier of Central and Eastern States, however, it is desirable to have a roof with even better insulating properties. This is done by the use of 3- by 6-inch decking with a cover of 25/32-inch insulating board sheathing, or by the use of an expanded foam plastic covered with ½-inch insulating board or tile on the underside of 2- by 6-inch decking. Figure 2-38 shows various materials and combinations of materials and their approximate "U" values.

The method of installing the $25/32$-inch insulating sheathing over the wood decking is relatively simple. The 4- by 8-foot sheets should be laid horizontally across the decking. Use 1¼- or 1½-inch roofing nails spaced 10 inches apart in rows 24 inches apart along the length.

Insulating board or tile in ½-inch thickness and the expanded foam insulation are installed with wallboard adhesive designed for these materials. Manufacturers normally recommend the type and method of application. With ½- by 12- by 12-inch tile, for example, a small amount of adhesive in each corner and the use of hand pressure as the tile is placed is one system which is often used. When tile is tongued and grooved, stapling is the usual method of installation. Larger sheets of ½-inch insulating board may require a combination of glue and some nailing. Expanded foam insulation is ordinarily installed with approved adhesives.

When 1-inch wood decking is used over the joist-beam system (fig. 2-33 A and B), the use of at least $25/32$-inch insulating sheathing over the boards and ½-inch insulating board or tile on the inside is normally recommended. A 1-inch thickness of

MATERIAL COMBINATIONS
(INCLUDING ROOFING)

APPROXIMATE TOTAL "U" VALUE

2 x 6 WOOD DECKING — 0.41

3 x 6 CEDAR DECKING — 0.22

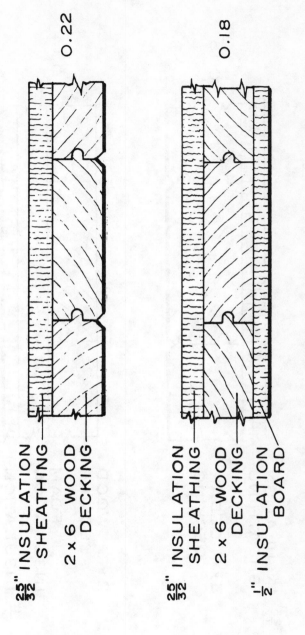

Figure 2-38. Insulating values of various materials and material combinations.

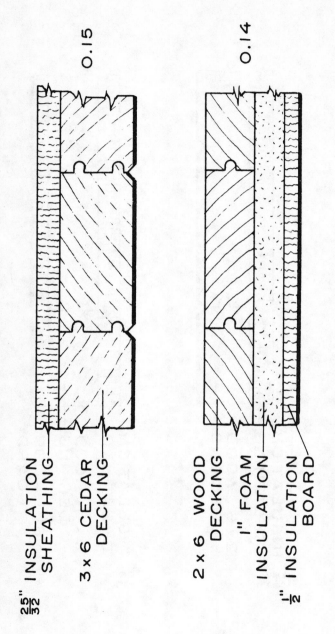

Figure 2-38. Cont'd.

expanded foam insulation under the ½-inch tile would provide even better insulation. When the inner face of the decking is to be covered, lower grade 2- by 6-inch decking is commonly used.

Trim For Low-Slope Roofs

Simple trim in the form of facia boards can be used at roof overhangs and at side and end walls. When 2- by 6-inch lookouts are used in the ceiling-beam roof, a 2- by 8-inch *facia* member is usually required to span the 48-inch spacing of the beams (fig.2-33*A*).In addition, a 1- by 2-inch facia molding may be added. Use two sixteenpenny galvanized nails in the ends of each lookout member. The facia molding may be nailed with six- or sevenpenny galvanized nails on 16-inch centers.

Trim for roofs with nominal 2- or 3-inch-thick wood decking can consist of a 1- by 4- or 1- by 6-inch member with a 1- by 2-inch facia molding at the side and end-wall overhangs (fig.2-39*A* and *B*). A 1- by 4-inch member can be used for 2-inch roof decking with or without the $25/_{32}$-inch insulating fiberboard. When 3-inch roof decking is used with the fiberboard, a 1- by 6-inch piece is generally required. Nail the facia and molding to the decking with eightpenny galvanized nails spaced about 16 inches apart. The roof deck is now ready for the roofing material.

Pitched Roof

A pitched-roof house is commonly framed by one of two methods: (a) With trussed rafters or (b) with conventional rafter and ceiling joist members. These framing methods are used most often for roof slopes of 4 in 12 and greater. The common W-truss (fig.2- 40*A*) for moderate spans requires less material than the joist and rafter system, as the members in the upper and lower chords are usually only 2 by 4 inches in size for spans

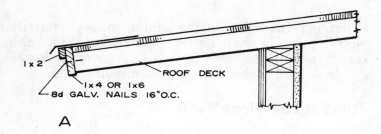

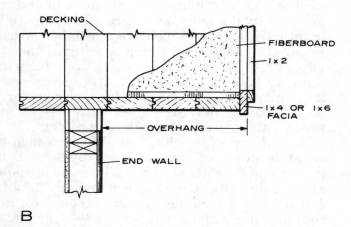

FIGURE 2-39. Facia for wood-deck roof. A, Sidewall overhang; B, end-wall overhang.

of 24 to 32 feet. The king-post truss (fig. 2-40*B*) for spans of 20 to 24 feet uses even less material than the W-truss, but is perhaps more suitable for light to moderate roof loads. Low-slope roof trusses usually require larger members. In addition to lowering material costs, the truss has the advantage of permitting freedom in location of interior partitions because only the sidewalls carry the ceiling and roof loads.

The roof sheathing, trim, roofing, interior ceiling finish, and type of ceiling insulation used do not vary a great deal between the truss and the conventional roof systems. For plywood or lumber sheathing, 24-inch spacing of trusses and rafters and joists is considered a normal maximum.

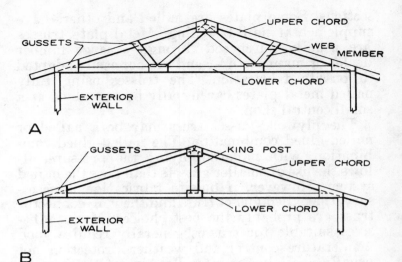

FIGURE 2-40. Trussed rafters. A, W-type truss; B, king-post truss.

Greater spacing can be used, but it usually requires a thicker roof sheathing and application of wood stripping on the undersides of the ceiling joists and trusses to furnish a support for ceiling finish. Thus, most W-trusses are designed for 24-inch spacing and joist-rafter construction for 24- or 16-inch spacing. Trusses generally require a higher grade dimension material than the joist and rafter roof. However, specific details of the roof construction are covered in the working drawings for each house.

Trussed Roof

The common truss or trussed rafter is most often fabricated in a central shop. While some are constructed at the job side, an enclosed building provides better control for their assembly. These trusses are fabricated in several ways. The three most common methods of fastening members together are with (a) metal truss plates, (b) plywood gussets, and (c) ring connectors.

The metal truss plates, with or without prongs, are fastened in place on each side of member intersections. Some plates are nailed and others have supplemental nail fastening. Metal-plate trusses are usually purchased through a large lumber dealer or manufacturer and are not easily adapted to on-site fabrication. The trusses using fully nailed metal plates can usually be assembled at a small central shop.

The plywood-gusset truss may be a nailed or nailed-glued combination. The nailed-glued combination, with nails supplying the pressure, allows the use of smaller gussets than does the nailed system. However, if on-site fabrication is necessary, the nailed gusset truss and the ring connector truss are probably the best choices. Many adhesives suitable for trusses generally require good temperature control and weather protection not usually available on site. The size of the gussets, the number of nails or other connectors, and other details for this type of roof are included in the working drawings for each house.

Completed trusses can be raised in place with a small mechanical lift on the top plates of exterior sidewalls. They can also be placed by hand over the exterior walls in an inverted position, and then rotated into an upright position. The top plates of the two sidewalls should be marked for the location of each set of trusses. Trusses are fastened to the outside walls and to 1- by 4- or 1- by 6-inch temporary horizontal braces used to space and aline them until the roof sheathing has been applied. Locate these braces near the ridge line.

Trusses can be fastened to the top wallplates by toenailing, but this is not always the most satisfactory method. The heel gusset in a plywood-gusset or metal plate truss is located at the wallplate and makes toenailing difficult. However, two tenpenny nails at each side of the truss can be used in nailing the lower chord to plate (fig. 2-41 A). Predrilling may be necessary to prevent splitting. A better system involves the use of a simple metal connector or bracket obtained from local lumber dealers. Brackets should be nailed to the

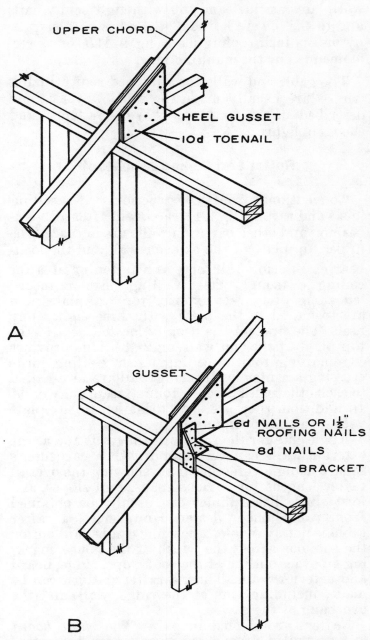

FIGURE 2-41. Fastening trusses to wallplate. A, Toenailing; B, metal bracket connector.

207

wallplates at sides and top with eightpenny nails and to the lower chords of the truss with sixpenny or 1½-inch roofing nails (fig. 2-41B) or as recommended by the manufacturer.

The gable-end walls for a pitched roof utilizing trusses are usually made the same way as those described in the section on "Wall Systems" and shown in figure 2-23.

Rafter and Ceiling Joist Roof

Conventional roof construction with ceiling joists and rafters (fig. 2-42) can begin after all loadbearing and other partition walls are in place. The upper top plate of the exterior wall and the loadbearing interior wall serve as a fastening area for ceiling joists and rafters. Ceiling joists are installed along premarket exterior top wallplates and are toenailed to the plate with three eightpenny nails. The first joist is usually located next to the top plate of the end wall (fig. 2-43). This provides edge-nailing for the ceiling finish. Ceiling joists crossing a center loadbearing wall are face-nailed to each other with three or four sixteenpenny nails. In addition, they are each toenailed to the plate with two eightpenny nails.

Angle cuts for the rafters at the *ridge* and at the exterior walls can be marked with a carpenter's square using a reference table showing the overall rafter lengths for various spans, roof slopes, and joist sizes. These tables can usually be obtained from your lumber dealer. However, if a rafter table is not available, a baseline can be laid out on the subfloor across the width of the house, marking an exact outline of the roof slope, ridge, board and exterior walls. Thus, a rafter pattern can be made, including cuts at the ridge, wall, and the overhang at the eaves.

Rafters are erected in pairs. The *ridge board* is first nailed to one rafter end with three tenpenny nails (fig. 2-44). The opposing rafter is then

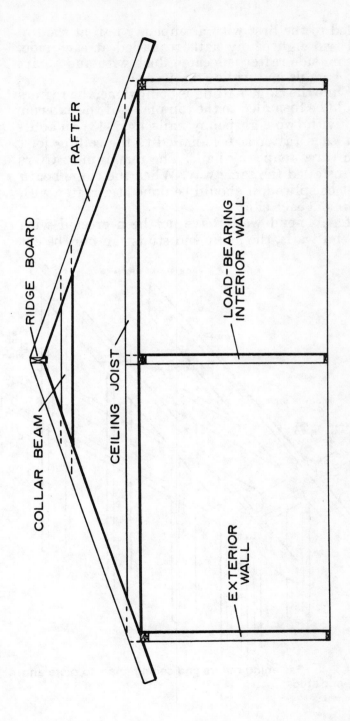

FIGURE 2-42. Rafter and ceiling joist roof framing.

nailed to the first with a tenpenny nail at the top and two eightpenny nails toenailed at each side. The outside rafter is located flush with and a part of the gable-end walls (fig. 2-43).

While the ridge nailing is being done, the rafters should be toenailed to the top plates of the exterior wall with two eightpenny nails (fig. 2-43). In addition, each rafter is face-nailed to the ceiling joist with three tenpenny nails. The remaining rafters are installed the same way. When the ridge board must be spliced, it should be done at a rafter with nailing at each side.

If gable-end walls have not been erected with the end walls, the gable-end studs can now be cut

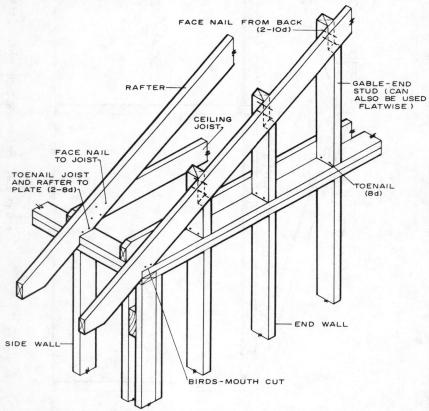

FIGURE 2-43. Fastening rafters and ceiling joists to plate and gable-end studs.

and nailed in place (fig. 2-43). Toenail the studs to the plate with eightpenny nails and face-nail to the end rafter from the inside with two tenpenny nails. In addition, the first or edge ceiling joist can be nailed to each gable-end stud with two tenpenny nails. Gable-end studs can also be used flatwise between the end rafter and top plate of the wall.

When the roof has a moderately low slope and the width of the house is 26 feet or greater, it is often desirable to nail a 1- by 6-inch *collar beam* to every second or third rafter (fig. 2-42) using four eightpenny nails at each end.

Framing for Flush Living-Dining Area Ceiling

A living-dining-kitchen group is often designed as one open area with a flush ceiling throughout. This makes the rooms appear much larger than they actually are. When trusses are used, there is no problem, because they span from one exterior wall to the other. However, if ceiling joists and

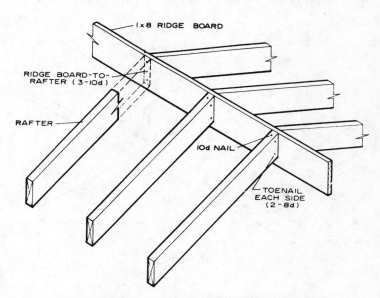

FIGURE 2-44. Fastening rafters at the ridge.

rafters are used, some type of *beam* is needed to support the interior ends of the ceiling joists. This can be done by using a flush beam, which spans from an interior cross wall to an exterior end wall. Joists are fastened to the beams by means of joist hangers (fig. 2-45). These hangers are nailed to the beam with eightpenny nails and to the joist with sixpenny nails or 1½-inch roofing nails. Hangers are perhaps most easily fastened by first nailing to the end of the joist before the joist is raised in place.

An alternate method of framing utilizes a wood bracket at each pair of ceiling joists tieing them to a beam which spans the open living-dining area (fig. 2-46). This beam is blocked up and fastened at each end at a height equal to the depth of the ceiling joists.

Roof Sheathing

Plywood or lumber roof sheathing is most commonly used for pitched roofs. Nominal 1-inch boards no wider than 8 inches can be used for trusses or rafters spaced not more than 24 inches on center. Sheathing (standard) and other grades

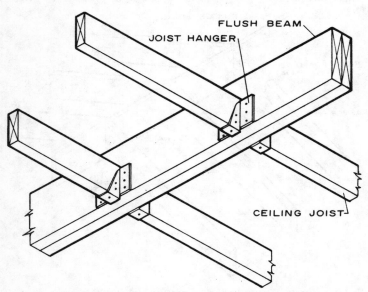

FIGURE 2-45. Flush beam with hoist hangers.

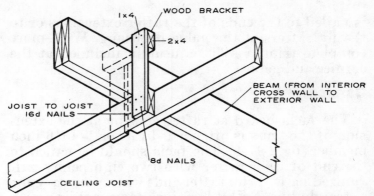

FIGURE 2-46. Framing for flush ceiling with wood brackets.

of plywood are marked for the allowable spacing of the rafters and trusses for each species and thickness used. For example, a "24/0" mark indicates it is satisfactory as roof sheathing for 24-inch spacing of roof members, but not satisfactory for subfloor.

Nominal 1-inch boards should be laid up without spacing and nailed to each rafter with two eightpenny nails. Plywood sheets should be laid across the roof members with staggered end joints. Use sixpenny nails for 3/8-inch and thinner plywood and eightpenny nails for 1/2-inch and thicker plywood. Space the nails 6 inches apart at the edges and 12 inches at intermediate fastening points.

When gable-end overhangs are used, extend the trim to the plywood or roofing boards when necessary before the 2- by 2- or 2- by 4-inch fly rafter (facia nailer) is nailed in place.

Roof Trim

Roof trim is installed before the roofing or shingles are applied. The cornice and gable (*rake*) trim for a pitched roof can be the same whether trusses or rafter-ceiling joist framing are used. In its simplest form, the trim consists of a facia board, sometimes with molding added. The facia is nailed to the ends of the rafter extensions or to the fly rafters at the gable overhang. With more complete trim, a soffit is usually included at the cornice and gables.

Cornice

The facia board at rafter ends or at the extension of the truss is often a 1- by 4- or 1- by 6-inch member (fig. 2-47A). The facia should be nailed to the end of each rafter with two eightpenny galvanized nails. Trim rafter ends when necessary for a straight line. Nail 1- by 2-inch facia molding with one eightpenny galvanized nail at rafter locations. In an open cornice, a frieze board is often used between the rafters, serving to terminate siding or siding-sheathing combinations at the rafter line (fig. 2-47A).

A simple closed cornice is shown in fig. 2-47 B. The soffit of plywood, hardboard, or other material is nailed directly to the underside of the rafter extensions. Blocking may be required between rafters at the wall line to serve as a nailing surface for the soffit. Use small galvanized nails in nailing the soffit to the rafters. When inlet attic ventilation is specified in the plans, it can be provided by a screened slot (fig. 2-47B), or by small separate ventilators.

When a horizontal closed cornice is used, *lookouts* are fastened to the ends of the rafter and to the wall (fig. 2-47C). They are face-nailed to the rafters and face- or toenailed to the studs at the wall. Use twelvepenny nails for the face-nailing and eightpenny nails for the toenailing.

Gable End

The gable-end trim may consist of a fly rafter, a facia board, and facia molding (fig. 2-48A). The 2- by 2- or 2- by 4-inch fly rafter is fastened by nailing through the roof sheathing. Depending on the thickness of the sheathing, use sixpenny

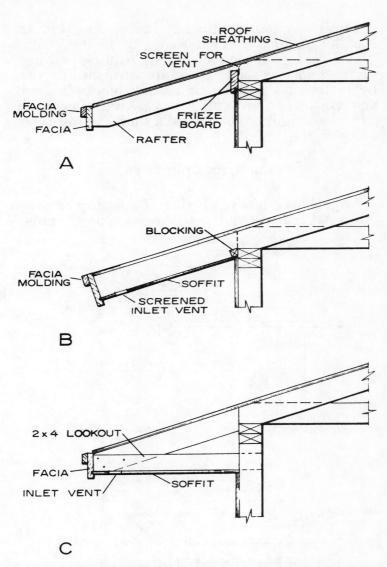

FIGURE 2-47. Cornice trim. A, Open cornice; B, sloped closed cornice; C, horizontal closed cornice.

or eightpenny nails spaced 12 inches apart. In this type of gable end, the amount of extension should be governed by the thickness of the roof sheathing. When nominal 1-inch boards or plywood thicker than ½ inch is used, the extension should generally be no more than 16 inches. For thinner sheathings, limit the extensions to 12 inches.

A closed gable-end overhang requires nailing surfaces for the soffit. These are furnished by the fly rafter and a nailer or nailing blocks located against the end wall (fig. 2-48*B*). An extension of 20 inches might be considered a limit for this type of overhang.

Framing for Chimneys

An inside chimney, whether of masonry or prefabricated, often requires that some type of fram-

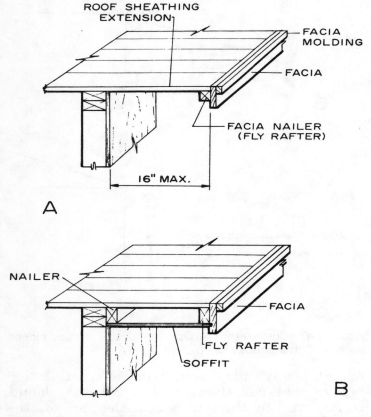

FIGURE 2-48. Gable-end trim. A, Open gable overhang; B, closed gable overhang.

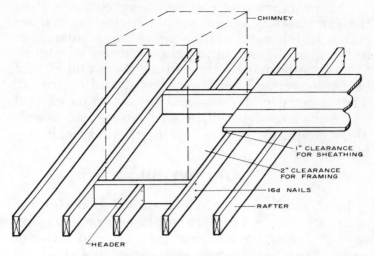

FIGURE 2-49. Roof framing at chimney.

ing be provided, where it extends through the roof. This may consist of simple headers between rafters below and above the chimney location, or require two additional rafter spaces (fig. 2-49). The chimney should have a 2-inch clearance from the framing members and 1 inch from roof sheathing. When nominal 2- or 3-inch wood decking is used, a small header can be used at each end of the decking at the chimney location for support.

Chimneys

Some type of chimney will be required for the heating unit, whether the home is heated by oil, gas, or solid fuel. It is normally erected before the roofing is laid but also can be installed after. Chimneys, either of masonry or prefabricated, should be structurally safe and provide sufficient draft for the heating unit and other utilities. Local building regulations often dictate the type to be used. A masonry chimney requires a stable foundation below the frostline and construction with acceptable brick or other masonry units. Some type of *flue lining* is included, together with a cleanout door at the base.

The prefabricated chimney may cost less than the full masonry chimney, considering both materials and labor, as well as providing a small saving in space. These chimneys are normally fastened to and supported by the ceiling joists and should be Underwriter Laboratory tested and approved. They are normally adapted to any type of fuel and come complete with roof flashing, cap assembly, mounting panel, piping, and chimney housing.

ROOF COVERINGS

Roof coverings should be installed soon after the cornice and rake trim are in place to provide protection for the remaining interior and exterior work. For the low-cost house, perhaps the most practical roof coverings are roll roofing or asphalt shingles for pitched roofs and roll roofing in double coverage or *built-up roof* for flat or very low-slope roofs. A good maintenance-free roof is important from the standpoints of protection and the additional cost involved in replacing a cheaper roof after only a few years.

Asphalt Shingles

Asphalt *shingles* may be used for roofs with slopes of 2 in 12 to 7 in 12 and steeper under certain conditions of installation. The most common shingle is perhaps the 3 in 1, which is a 3-tab strip, 12 by 36 inches in size. The basic weight may vary somewhat, but the 235-pound (per square of 3 bundles) is now considered minimum. However, many roofs with 210-pound shingles are giving satisfactory service. A small gable roof house uses about 10 squares of shingles, so use of a better shingle would mean about $10 to $20 more per house. Cost of application would be the same.

Installation

Underlay.—A single underlay of 15-pound saturated felt is used under the shingles for roof

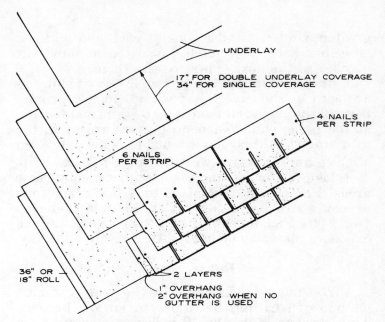

FIGURE 2-50. Installing asphalt shingles.

slopes of 4 in 12 to 7 in 12. A double underlay (double coverage) is required for slopes of 2 in 12 to 4 in 12. Roof slopes over 7 in 12 usually require no underlay. For single underlay, start at the eave line with the 15-pound felt, roll across the roof, and nail or staple the felt in place as required. Allow a 2-inch head lap and install the second strip. This leaves a 34-inch exposure for the standard 36-inch-width rolls. Continue in this manner.

A double underlay can be started with two layers at the eave line, flush with the facia board or molding. The second and remaining strips have 19-inch head laps with 17-inch exposures (fig. 2-50) Cover the entire roof in this manner, making sure that all surfaces have double coverage. Use only enough staples or roofing nails to hold the underlay in place. Underlay is normally not required for wood shingles.

Shingles.—Asphalt tab shingles are fastened in place with $3/4$- or $7/8$-inch galvanized roofing nails or with staples, using at least four on each strip (fig. 2-50). Some roofers use six for each strip for greater wind resistance; one at each end and one at each side of each notch. Locate them above the notches so the next course covers them.

A starter strip and one or two layers of shingles are used at the eave line with a 1-inch overhang beyond the facia trim and ½- to ¾-inch extension at the gable end. When no gutters are used, the overhang should be about 2 inches. This will form a curve during warm weather for a natural drip. Metal edging or flashing is sometimes used at these areas. For slopes of 2 in 12 to 4 in 12, a 5-inch exposure can be used with the double underlay (fig. 2-50). For slopes of 4 in 12 and over, a 5-inch exposure may also be used with a single underlay.

Ridge

A *Boston ridge* is perhaps the most common method of treating the ridge portion of the roof. This consists of 12- by 12-inch sections cut from the 12- by 36-inch shingle strips. They are bent slightly and used in lap fashion over the ridge with a 5-inch exposure distance (fig. 2-51). In cold weather, be careful that the sections do not crack in bending. The nails used at each side are covered

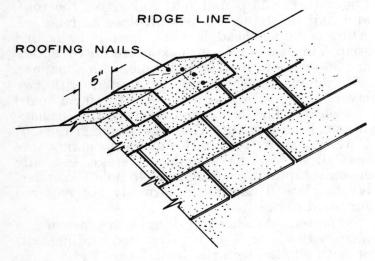

FIGURE 2-51. Boston ridge.

by the lap of the next section. For a positive seal, use a small spot of asphalt cement under each exposed edge.

Roll Roofing

When cost is a factor in construction of a house, the use of mineral-surfaced roll roofing might be considered. While this type of roofing will not be as attractive as an asphalt shingle roof and perhaps not as durable, it may cost up to 15 percent less for a small house than standard asphalt shingles.

Roll roofing (65 pounds minimum weight in one-half lap rolls with a mineral surface) should be used over a double underlay coverage. Use a starter strip or a half-roll at the eave line with a 1-inch overhang and nail in place 3 to 4 inches above the edge of the facia (fig. 2-52). When *gutters* are not included initially, use a 2-inch extension

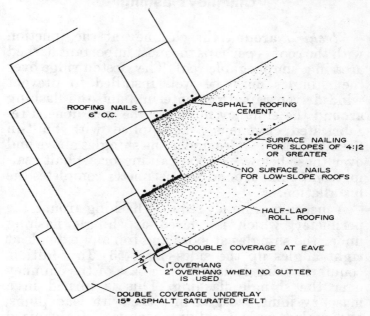

FIGURE 2-52. Installing roll roofing.

to form a drip edge. Space roofing nails about 6 inches apart. Surface nailing can be used when roof slopes are 4 in 12 and greater.

The second (full) roll is now placed along the eave line over a ribbon of asphalt roofing cement or lap-joint material. In low slopes, nailing is done above the lap, cement applied, and the next roll positioned so that the nails are covered. Edge overhang should be about $\frac{1}{2}$ to $\frac{3}{4}$ inch at the gable ends. When vertical lap joints are required, nail the first edge, then use asphalt adhesive under a minimum 6-inch overlap. Use a sufficient amount of adhesive or lap-joint material to insure a tight joint. On steep slopes, surface nailing along the vertical edge is acceptable. The ridge can be finished with a Boston-type covering or by 12-inch-wide strips of the roll roofing, using at least 6 inches on each side.

Chimney Flashing

Flashing around the chimney at the junction with the roof is perhaps the most important flashed area in a simple gable roof. The Boston ridge over the shingles must be well installed to prevent wind-driven rain from entering, and the flashing around the chimney must also be well done. Prefabricated chimneys are supplied with built-in flashing which slides under the shingles above and over those below. A good calking or asphalt sealing compound around the perimeter completes the installation.

A masonry chimney requires flashing around the perimeter, which is placed as shingle flashing under the shingles at sides and top and extends at right angles up the sides (fig. 2-53). In addition, counterflashing is used on the base of the chimney over the shingle flashing. This is turned in a masonry joint, wedged in place with lead plugs, and sealed with a calking material. Galvanized sheet metal, aluminum, and terneplate (coated sheet iron or steel) are the most common types

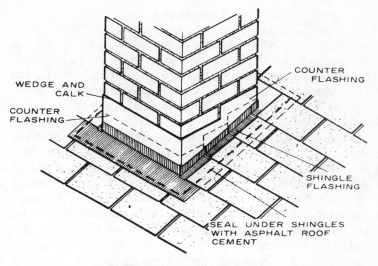

FIGURE 2-53. Chimney flashing.

used for flashing around the chimney. If they are not rust-resistant, they should be given a coat or two of good metal paint.

EXTERIOR WALL COVERINGS

Exterior coverings used over the wall framing commonly consist of a sheathing material followed by some type of finish siding. However, sheathing-siding materials (panel siding) serve as both sheathing and finish material. These materials are most often plywoods or hardboards. While they are somewhat higher in price than conventional sheathing alone, they make it possible to use only a single exterior covering material. Low-cost sheathing materials can be covered with various types of siding—from spaced vertical boards over plywood sheathing to horizontal bevel siding over fiberboard, plywood, or other types of sheathing. All combinations should be studied so that cost, utility, and appearance are considered in the selection. The working drawings of the house indicate the most suitable siding materials.

Sheathing

In a low-cost house, it is advisable to use a sheathing or a panel-siding material which will provide resistance to racking and thus eliminate the need for diagonal corner bracing on the stud wall. Notching studs and installing bracing can add substantially to labor cost. When siding material does not provide this rigidity and strength, some type of sheathing should be used. Materials which provide resistance to racking are: (a) Diagonal board sheathing, (b) structural insulation board (fiberboard) sheathing in $25/32$-inch regular density or $1/2$-inch intermediate fiberboard or nail-base fiberboard sheathing for direct application of shingles, and (c) plywood. The fiberboard and plywood sheathing must be applied vertically in 4- by 8-foot or longer sheets with edge and center nailing to provide the needed racking resistance. Horizontal wood boards may also be used for sheathing but require some type of corner bracing.

Diagonal Boards

Diagonal wood sheathing should have a nominal thickness of $5/8$ inch (resawn). Edges can be square, *shiplapped*, or tongued-and-grooved. Widths up to 10 inches are satisfactory. Sheathing should be applied at as near a 45° angle as possible as shown in figure 2-54. Use three eightpenny nails for 6- and 8-inch-wide boards and four eightpenny nails for the 10-inch widths. Also provide nailing along the floor framing or beam faces. Butt joints should be made over a stud unless the sheathing is end and side matched. Depending on the type of siding used, sheathing should normally be carried down over the outside floor framing members. This provides an excellent tie between wall and floor framing.

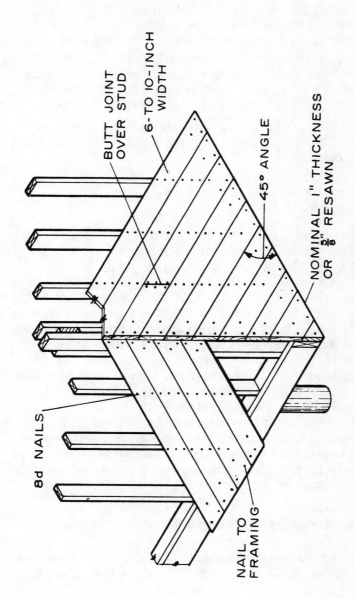

FIGURE 2-54. Diagonal board sheathing.

Structural Insulating Board

Structural *insulating board* sheathing (fiberboard type) in 4-foot-wide sheets and in $25/32$-inch regular-density or $1/2$-inch intermediate fiberboard grades provides the required rigidity without bracing. It must be applied vertically in 8-foot and longer sheets with edge and center nailing (fig. 2-55). Nails should be spaced 3 inches apart along the edges and 6 inches apart at intermediate supports. Use $1 3/4$-inch roofing nails for the $25/32$-inch sheathing and $1 1/2$-inch nails for the $1/2$-inch sheathing. Vertical joints should be made over studs. Siding is normally required over this type of sheathing.

Plywood

Plywood sheathing also requires vertical application of 4-foot-wide by 8-foot or longer sheets (fig. 2-55). Standard (sheathing) grade plywood is normally used for this purpose. Use $5/16$ inch minimum thickness for 16-inch stud spacing and $3/8$ inch for 24-inch stud spacing. Nails are spaced 6 inches apart at the edges and 12 inches apart at intermediate studs. Use sixpenny nails for $5/16$-inch $3/8$-inch-thick plywood. Because the plywood sheathing in $5/16$- or $3/8$-inch sheets provides the necessary strength and rigidity, almost any type of siding can be applied over it.

Plywood and hardboard are also used as a single covering material without sheathing, but grades, thickness, and types vary from normal sheathing requirements. This phase of wall construction will be covered in the following section.

Sheathing-Siding Materials—Panel Siding

Large sheet materials for exterior coverage (*panel siding*) can be used alone and serve both as

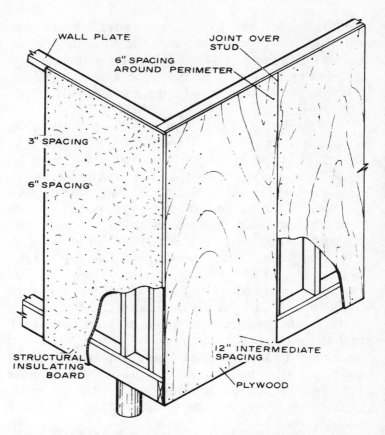

Figure 2-55. Sheathing with insulating board or plywood (vertical application).

sheathing and siding. Plywood, hardboard, and exterior particleboard in their various forms are perhaps the most popular materials used for this purpose. The proper type and size of plywood and hardboard sheets with adequate nailing eliminate the need for bracing. Particleboard requires corner bracing.

These materials are quite reasonable in price, and plywood, for example, can be obtained in grooved, rough-sawn, embossed, and other surface variations as well as in a paper-overlay form. Hardboard can also be obtained in a number of surface variations. The plywood surfaces are most suitable for pigmented stain finishes in various colors. The medium-density, paper-overlay plywoods are an excellent base for exterior paints. Plywoods used for panel siding are normally exterior grades.

The thickness of plywood used for siding varies with the stud spacing. Grooved plywood, such as the "1–11" type, is normally $5/8$ inch thick with $3/8$- by $1/4$-inch-deep grooves spaced 4 or 6 inches apart. This plywood is used when studs are spaced a maximum of 16 inches on center. Ungrooved plywoods should be at least $3/8$ inch thick for 16-inch stud spacing and $1/2$ inch thick for 24-inch stud spacing. Plywood panel siding should be nailed around the perimeter and at each intermediate stud. Use sixpenny galvanized siding or other rust-resistant nails for the $3/8$-inch plywood and eightpenny for $1/2$-inch and thicker plywood and space 7 to 8 inches apart. Hardboard must be $1/4$ inch thick and used over 16-inch stud spacing. Exterior particleboard with corner bracing should be $5/8$ inch thick for 16-inch stud spacing and $3/4$ inch thick for 24-inch stud spacing. Space nails 6 inches apart around the edges and 8 inches apart at intermediate studs.

The vertical joint treatment over the stud may consist of a shiplap joint as in the "1–11" paneling (fig. 2-56A). This joint is nailed at each side after treating with a water-repellent preservative. When a square-edge butt joint is used, a sealant calk should be used at the joint (fig. 2-56B).

A square-edge butt joint may be covered with battens, which can also be placed over each stud as a decorative variation (fig. 2-56C). Joints should be calked and the batten nailed over the joint with eightpenny galvanized nails spaced 12 inches apart. Nominal 1- by 2-inch battens are commonly used.

A good detail for this type of siding at gable ends consists of extending the bottom plate of the

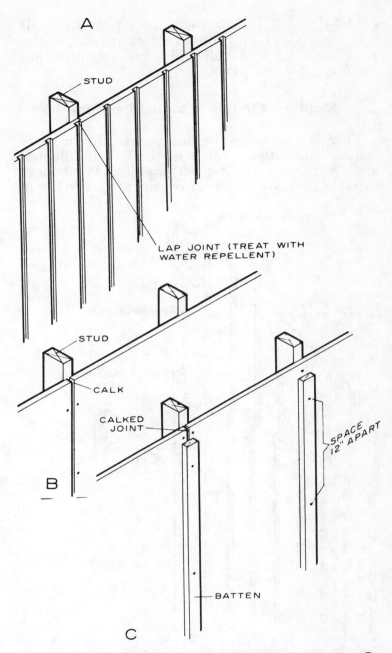

FIGURE 2-56. Joint treatment for panel siding. A, Lap joint; B, calked butt joint; C, butt joint with batten.

229

gable ⅝ to ¾ inch beyond the top of the wall below (fig. 2-57). This allows a termination of the panel at the lower wall and a good drip section for the gable-end panel.

Siding—With and Without Sheathing

There are a number of sidings, mainly for horizontal application, which might be suitable for walls with or without sheathing. The types most suitable for use over sheathing are: (a) The lower

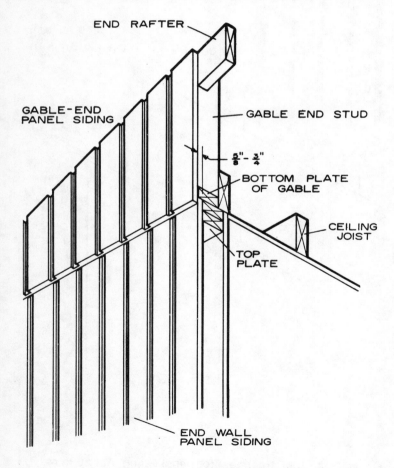

FIGURE 2.57. Panel siding at gable end.

cost lap sidings of wood or hardboard, (b) wood or other type shingles with single or double coursing, (c) vertical boards, and (d) several nonwood materials. Initial cost and maintenance should be the criteria in the selection. *Drop siding* and nominal 1-inch paneling materials can be used without sheathing under certain conditions. However, such sidings require (a) a rigidly braced wall at each corner and (b) a waterproof paper over the studs before application of the siding.

Application

Bevel siding.—When siding is used over sheathing, window and door frames are normally in-

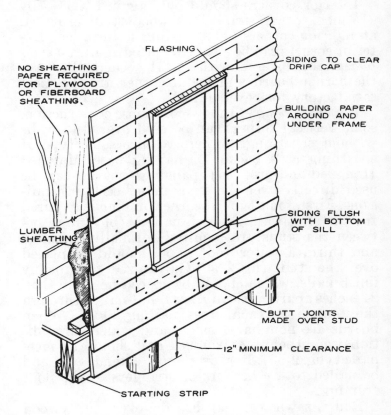

FIGURE 2-58. Installing bevel siding.

stalled first. This process will be discussed in the next section, "Exterior Frames." The exposed face of sidings such as *bevel siding* in ½- by 6-inch, ½- by 8-inch, or other sizes should be adjusted so that the butt edges coincide with the bottom of the sill and the top of the *drip cap* of window frames (fig. 2-58).Use a sevenpenny galvanized siding nail or other corrosion-resistant nail at each stud crossing for the ½-in-thick siding. The nail should be located so as to clear the top edge of the siding course below. Butt joints should be made over a stud. Other horizontal sidings over sheathing should be installed in a similar manner. Nonwood sheathings require nailing into each stud.

Interior corners should butt against a 1½- by 1½-inch corner strip. This wood strip is nailed at interior corners before siding is installed. Exterior corners can be mitered, butted against *corner boards*, or covered with metal corners. Perhaps the corner board and metal corner are the most satisfactory for bevel siding.

Vertical siding.—In low-cost house construction, some vertical sidings can be used over stud walls without sheathings; others require some type of sheathing as a backer or nailing base. Matched (tongued-and-grooved) paneling boards can be used directly over the studs under certain conditions. First, some type of corner bracing is required for the stud wall. Second, nailers (blocking) between the studs are required for nailing points, and third, a waterproof paper should be placed over the studs (fig. 2-59).Galvanized sevenpenny finish nails, which should be spaced no more than 24 inches apart vertically, are blind-nailed through the tongue at each cross nailing block. When boards are nominal 6 inches and wider, an additional eightpenny galvanized nail should be face-nailed (fig. 2-59). Boards should extend over and be nailed to the headers or stringers of the floor framing.

Rough-sawn vertical boards over a plywood backing provide a very acceptable finish. The ply-

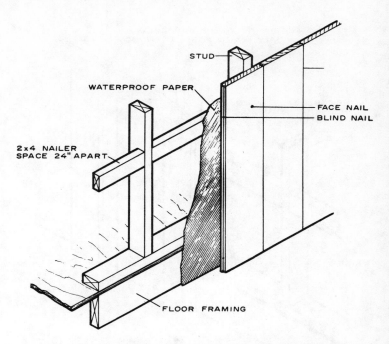

FIGURE 2-59. Vertical paneling boards over studs.

wood should be an exterior grade or sheathing grade (standard) with exterior glue. It should be ½ inch thick or ⁵⁄₁₆ inch thick with nailing blocks between studs (fig. 60). Rough-sawn boards 4 to 8 inches wide surfaced on one side can be spaced and nailed to the top and bottom wallplates and the floor framing members and to the nailing blocks (fig. 2-60). Use the surfaced side toward the plywood. A choice in the widths and spacings of boards allows an interesting variation between houses.

FRAMING DETAILS

Windows and Doors

Window and exterior door units, which include the frames as well as the *sash* or doors, are generally assembled in a manufacturing plant and arrive at the building site ready for installation.

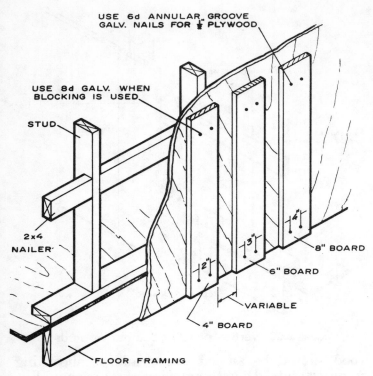

FIGURE 2-60. Vertical boards over plywood.

Doors may require fitting, however. Simple jambs and sill units for awning or hopper window sash can be made in a small shop with the proper equipment, the wood treated with a water-repellent preservative, and sash fitted and prehung. Only the sash would need to be purchased. However, this system of fabrication is practical only for the simplest units and only when a large number of the same type are required. Thus, for double-hung and sill units for awning or hopper window sash desirable to select the lower cost standard-size units and use a smaller number of windows. A substantial saving can be made, for example, by using one large double-hung window rather than two smaller ones.

Window frames are generally made with nominal 1-inch jambs and 2-inch sills. Sash for the most part are 1⅜ inches thick. Exterior door frames are made from 1½- to 1¾-inch stock. Exterior doors are 1¾ inches thick and the most common are the flush and the panel types.

As a general rule, the amount of natural light provided by the glass area in all rooms (except the kitchen) should be about 10 percent of the floor area. The kitchen can have natural or artificial light, but when an operating window is not available, ventilation should be provided. A bathroom usually is subject to the same requirements. From the standpoint of safety, houses should have two exterior doors. Local regulations often specify any variations of these requirements. The main exterior door should be 3 feet wide and at least 6 feet 6 inches high; 6 feet 8 inches is a normal standard height for exterior doors. The service or rear door should be at least 2 feet 6 inches wide; 2 feet 8 inches is the usual width. These details are covered in the working drawings for each house.

Types of Windows and Doors

Perhaps the most common type of window used in houses is the double-hung unit (fig. 2-61). It can be obtained in a number of sizes, is easily *weatherstripped*, and can be supplied with storms or screens. Frames are usually supplied with pre-fitted sash and the exterior casing and drip cap in place.

Another type of window, which is quite reasonable in cost and perhaps the one most adaptable to small shop fabrication in a simple form, is the "awning" or "hopper" type (fig. 2-62).

Other windows, such as the *casement sash* and sliding units, are also available but generally their cost is somewhat greater than the two types described. The fixed or stationary sash may consist of a simple frame with the sash permanently fastened in place. The frame for the awning window would be suitable for this type of sash (fig. 2-62).

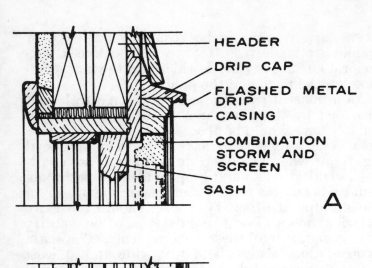
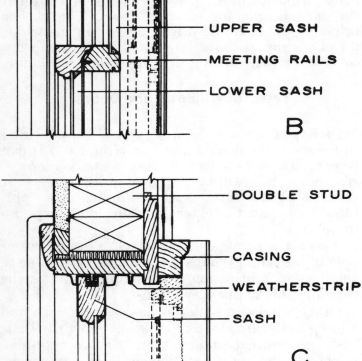

Figure 2-61. Double-hung window unit. Cross sections: A, head jamb; B, meeting rails; C, side jamb; D, sill.

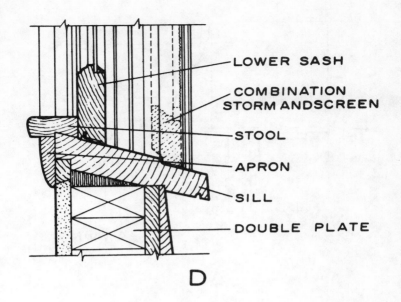

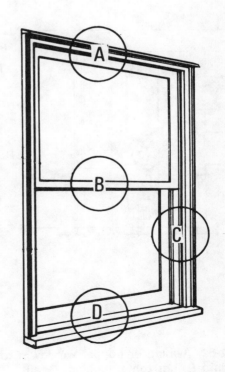

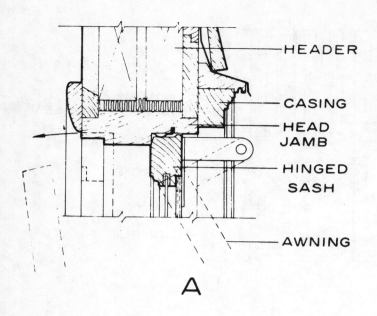
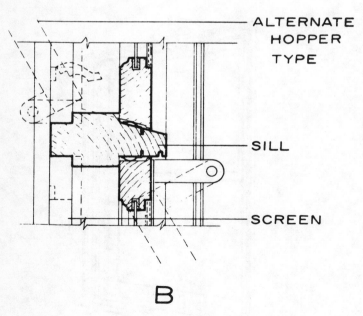

FIGURE 2-62. Awning or hopper window. Cross sections: A, head jamb; B, horizontal mullion; C, sill.

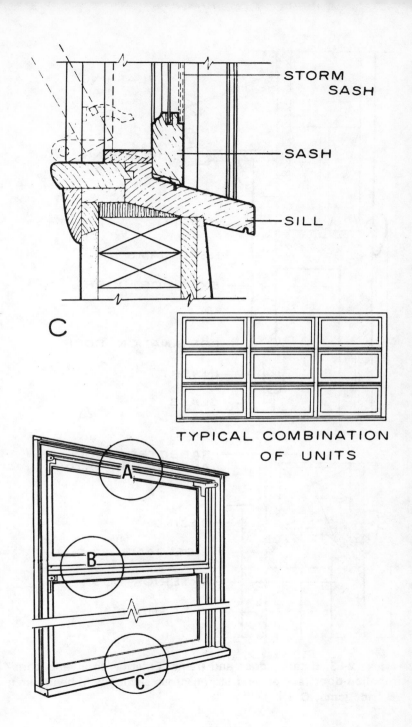
TYPICAL COMBINATION OF UNITS

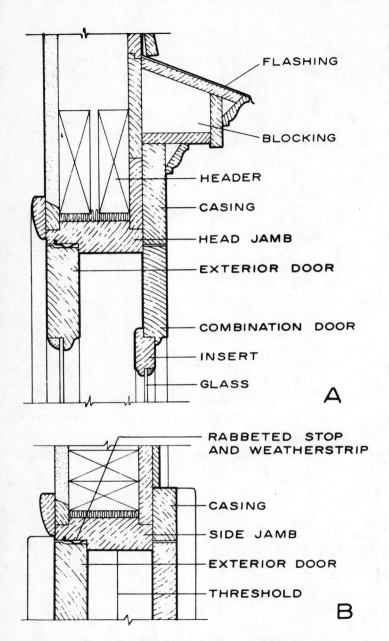

FIGURE 2-63. Exterior door and frame. Exterior door and combination-door (screen and storm) cross sections: A, head jamb; B, side jamb; C, sill.

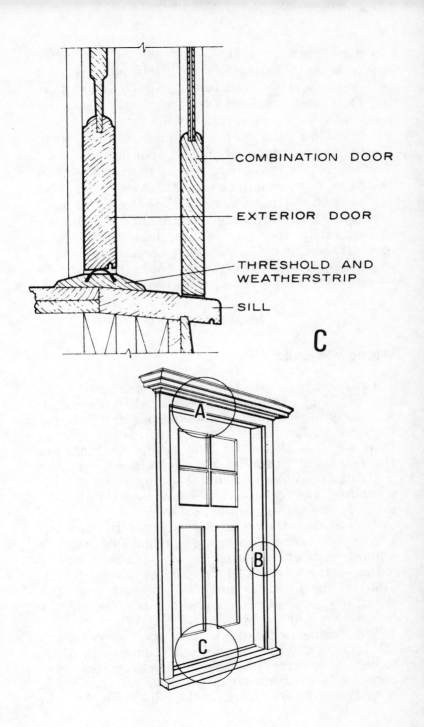

Door frames are also supplied with exterior side and head casing and a hardwood sill or a softwood sill with a reinforced edge (fig.2-63). Perhaps the most practical exterior door, considering cost and performance, is the panel type. A number of styles and patterns are available, most of them featuring some type of glazed opening (fig.2-64). The solid-core flush door, usually more costly than the panel type, should be used for exteriors in most central and northern areas of the country in preference to the hollow-core type. A hollow-core door is ordinarily for interior use, because it warps excessively during the heating season when used on the outside. However, it would probably be satisfactory for exterior use in the southern areas.

Installation

Window Frames

Preassembled window frames are easily installed. They are made to be placed in the rough wall openings from the outside and are fastened in place with nails. When a panel siding is used in place of a sheathing and siding combination, the frames are usually installed after the siding is in place. When horizontal siding is used with sheathing, the frames are fastened over the sheathing and the siding applied later.

To insure a water- and windproof installation for a panel-siding exterior, a ribbon of calking sealant (rubber or similar base) is placed over the siding at the location of side and head casing (fig. 2-65). When a siding material is used over the sheathing, strips of 15-pound asphalt felt should be used around the opening (fig. 2-58).

The frame is placed in the opening over the calking sealant (preferably with the sash in place to keep it square), and the sill leveled with a carpenter's level. Shims can be used on the inside if necessary. After leveling the sill, check the side

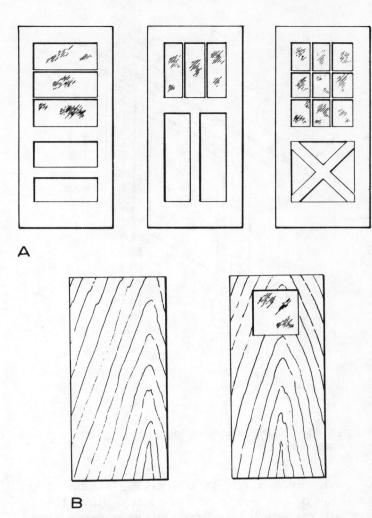

FIGURE 2-64. Exterior doors. A, Panel type; B, flush type.

casing and jamb with the level and square. Now nail the frame in place using tenpenny galvanized nails through the casing and into the side studs and the header over the window (fig. 2-66). While nailing, slide the sash up and down to see that they work freely. The nails should be spaced about 12 inches apart and both side and head casing fastened in the same manner. Other types of window units are installed similarly. When a panel siding

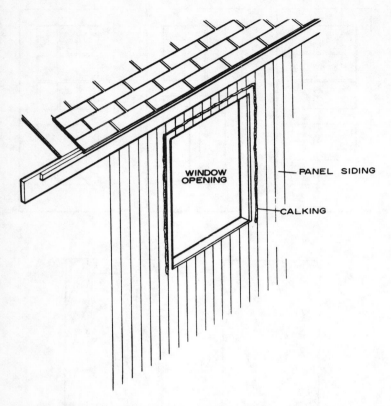

FIGURE 2-65. Calking around window opening before installing frame.

is used, place a ribbon of calking sealer at the junction of the siding and the sill. Follow this by installing a small molding such as quarter-round.

Door Frames

Door frames are also fastened over panel siding by nailing through the side and head casing. The header and the joists must first be cut and trimmed (fig. 2-67). Use a ribbon of calking sealer under the casing. The top of the sill should be the same height as the finish floor so that the *threshold* can be installed over the joint. The sill should be shimmed when necessary to have full bearing on

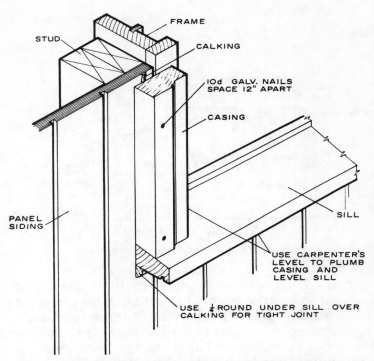

FIGURE 2-66. Installation of double-hung window frame.

the floor framing. A *quarter-round* molding in combination with calking when necessary for a tight windproof joint should be used under the door sill when a panel siding or other single exterior covering is used. When joists are parallel to the plane of the door, headers and a short support member are necessary at the edge of the sill (fig. 2-67). The threshold is installed after the finish floor has been laid.

PLUMBING AND OTHER UTILITIES

In cold northern areas, installing plumbing in a basementless house requires a little more care than in a house with a basement. Some protection for the drain and supply piping in the form of an insulated box from subfloor to the ground is re-

245

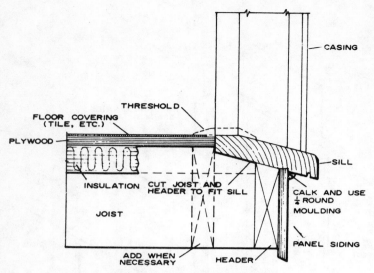

FIGURE 2-67. Door installation at sill.

quired. Thus, supply and drain lines should be located so as to eliminate long runs. Most plans for low-cost houses will back the kitchen and bath against a common utility wall so that all plumbing lines can be concentrated there for lower cost installation. In those houses which do not include all the facilities initially, plumbing lines should be roughed in so that connections can be made at a later date with little trouble.

Floor framing should be arranged so that little or no cutting of joists is required to install closet bends and other drainage and supply lines. This may require the use of a small header, for example, to frame out for the connections.

Plumbing Stack Vents

The utility wall between the kitchen and bath should be constructed so that connections can be made easily. This is usually done by using a nominal 2- by 6-inch plate and placing the studs flatwise at each side (fig. 2-68A). This will provide the needed wall thickness for the bell of a 4-inch cast-iron soil pipe, which is larger than the thickness

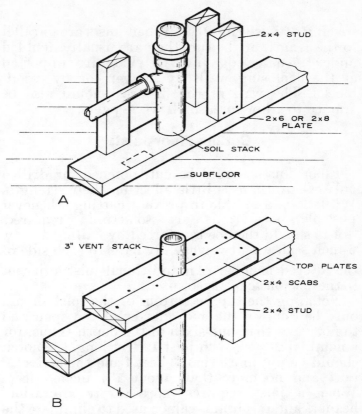

FIGURE 2-68. Framing for vent stack. A, 4-inch soil pipe; B, 3-inch stack vent.

of a 2- by 4-inch stud wall. It is also possible to furr out several studs to a 6-inch width at the *soil stack*, rather than thickening the entire wall. In areas where building regulations permit the use of a 3-inch vent pipe, a 2- by 4-inch stud wall may be used, but it requires reinforcing scabs at the top plate (fig. 2-68*B*). Use twelvepenny nails to fasten the scabs.

Bathtub Framing

The floor joists in the bathroom which support the tub or shower should be arranged so that no cutting is necessary in connecting the drainpipe. This usually requires only a small adjustment in spacing of joists(fig. 2-69). When joists are parallel to the length of the tub, they are usually doubled under the outer edge (fig.2-69). Tubs are supported at the enclosing walls by hangers or by woodblocks. The wall at the fixtures should also be framed to allow for a small access panel.

Cutting Floor Joists

Floor joists should be cut, notched, or drilled only where there is little effect on their strength. While it is desirable to prevent cutting whenever possible, alterations are sometimes required. Joists should then be reinforced by nailing a 2- by 6-inch scab with twelvepenny nails to each side of the altered member. An additional joist adjacent to the cut joist can also be used.

Notching the top or bottom of the joist should only be done in the end quarter of the span and to not more than one-sixth of the depth. Thus, for a nominal 2- by 8-inch joist 12 feet long, the notch should be not more than 3 feet from the end support and no more than about 1¾ inches deep. When a joist requires more severe alteration, headers and tail beams can be used to eliminate the need for cutting (fig. 2-70). Proper planning will minimize the need for altering joists.

When necessary, holes may be bored in joists if the diameters are no greater than 2 inches and the edges of the holes are not less than 2½ inches from the top or bottom edges (fig. 2-71). This usually limits a 2-inch-diameter hole to joists of nominal 2- by 8-inch size and larger.

Wiring

Wiring should be installed to comply with the National Electric Code or local building requirements. House wiring for electrical services is usually started sometime after the house has been enclosed. The initial phase of wiring is termed "roughing in" and includes installing outlet and

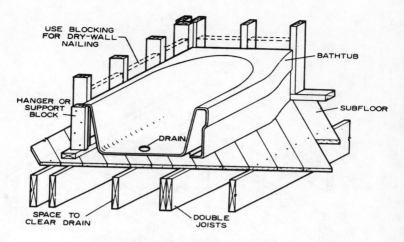

FIGURE 2-69. Framing for bathtub.

switch boxes and the connecting cable, with wire in the boxes ready for connecting. This work is done before the insulation is placed in the wall and before application of the dry-wall finish.

Framing changes for wiring are usually minor and consist only of holes drilled in the studs for the flexible *conduit*. Wall switches at doors should be located at the side of the doubled studs so that no cutting is necessary. They should be 48 to 54 inches above the floor.

THERMAL INSULATION

Thermal insulation is used in a house to minimize heat loss during the heating season and to reduce the inflow of heat during hot weather. Resistance to the passage of warm air is provided by materials used in wall, ceiling, and floor construction.

In constructing a crawl-space house, other factors must be considered in addition to providing insulation, such as: (a) Use of a *vapor barrier* with the insulation; (b) protection from ground moisture by the use of a vapor barrier ground cover (especially true if the crawl space is enclosed with a full foundation or skirt boards); and (c)

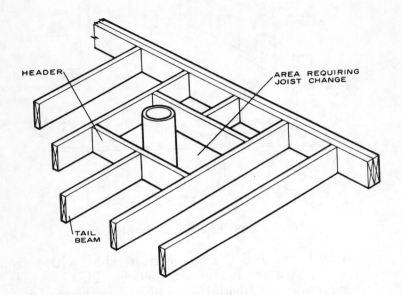

FIGURE 2-70. Headers for joists to eliminate cutting.

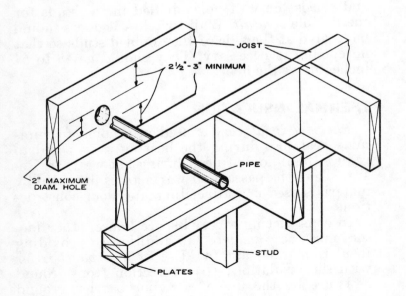

FIGURE 2-71. Boring holes in joists.

use of both attic and crawl-space ventilation when required. Insulation and vapor barriers are discussed here. Ventilation is covered in the next section.

The amount of insulation used in the walls, floor, and ceiling usually depends on what section of the country the house is located in. Low-cost houses constructed in the northern tier of States where winters are quite severe should have, when possible, at least a 2-inch blanket insulation or equivalent in the walls and 4 inches of fill or batt insulation above the ceiling. In a crawl-space house, a 4-inch friction-type batt or other type of insulation would normally be satisfactory for the floor.

For houses in the Central States with moderate winters, 1 inch in the walls, 4 inches in the ceilings, and 1 or 2 inches of insulation in the floor should be satisfactory. In the South, for economy, floor and wall insulation could be eliminated and the ceiling insulation reduced to a 2-inch blanket or batt. However, a 1-inch blanket in the walls and floors would help to provide desirable comfort during the hot summer months.

Types of Insulation

Commercial insulating materials which are most practical in the construction of low-cost houses are: (a) Flexible types in blanket and batt forms, (b) loose-fill types, and (c) *rigid insulation* such as building boards or insulation boards. Others include reflective insulations, expanded plastic foams, and the like.

The common types of blanket and batt insulation, as well as loose-fill types, are made from wood and cotton fiber and mineral wool processed from rock, slag, or glass. Insulating or building board may be made of wood or cane fibers, of glass fibers, and of expanded foam.

In comparing the relative insulating values of various materials, a 1-inch thickness of typical blanket insulation is about equivalent to (a) 1½ inches of insulating board, (b) 3 inches of wood, or (c) 18 inches of common brick. Thus, when practical, the use of even a small amount of thermal insulation is good practice.

The insulating values of several types of flexible insulation do not vary a great deal. Most loose-fill insulations have slightly lower insulating values than the same material in flexible form. However, fill insulations, such as *vermiculite*, have less than 60 percent the value of common flexible insulations. Most lower density sheathing or structural insulating boards have better insulating properties than vermiculite.

Vapor Barriers

Vapor barriers are often a part of blanket or batt insulation, but they may also be a separate material which resists the movement of water vapor to cold or exterior surfaces. They should be placed as close as possible to the warm side of all exposed walls, floors, and ceiling. When used as ground covers in crawl spaces, they resist the movement of soil moisture to exposed wood members. In walls, they eliminate or minimize condensation problems which can cause paint peeling. In ceilings they can, with good *attic ventilation*, prevent moisture problems in attic spaces.

Vapor barriers commonly consist of: (a) Papers with a coating or lamination of an asphalt material; (b) plastic films; (c) aluminum or other metal foils; and (d) various paint coatings. Most blanket and batt insulations are supplied with a laminated paper or an aluminum foil vapor barrier. Friction-type batts usually have no vapor barriers. For such insulation, the vapor barrier should be added after the insulation is in place. Vapor barriers should generally be a part of all insulating processes.

Vapor barriers are usually classed by their "*perm*" value, which is a rate of water-vapor movement through a material. The lower this value, the greater the resistance of the barrier. A perm rating of 0.50 or less is considered satisfactory for vapor barriers. Two-mil (0.002-inch-thick) polyethylene film, for example, has a perm rating of about 0.25.

When the crawl space is enclosed during or after construction of the house, or the soil under the house is quite damp, a *soil cover* should be used. This vapor barrier should have a perm value of 1.0 or less and should be laid over the ground, using about a 4-inch lap along edges and ends. Stones or half-sections of brick can be used at the laps and around the perimeter to hold the material in place. The ground should be leveled before placing the cover. Materials such as polyethylene, roll roofing, and asphalt laminated barriers are satisfactory for ground covers.

Where and How to Insulate

Insulation in some form should be used at all exterior walls, floors, and at the ceiling as a separate material or as a part of the house structure in most climates. The recommended thicknesses for various locations, given at the beginning of this chapter, can be used as a guide in insulating the house.

Floors

Blanket insulation in 1-inch thickness can be installed under a tongued-and-grooved plywood subfloor as previously described and illustrated in fig. |2-14. However, this type is most commonly applied between the joists. In applying insulation, be sure that the vapor barrier faces up, against the bottom of the plywood.

The use of friction or other types of insulating batts between joists has been shown in fig. 2-18. This insulation can be installed any time after the house has been enclosed and roofing installed. When friction-type insulation without a vapor barrier is used, a separate vapor barrier should be placed over the joists before the subfloor is nailed in place. Laminated or foil-backed kraft paper barriers or plastic films can be used. They should be lapped 4 to 6 inches and stapled only enough to hold the barriers in place until the subfloor is installed.

When batts are not the friction type, they usually require some support in addition to the adhesive shown in fig. 2-18. This support can be supplied simply by the use of small, 3/16- by 3/4-inch or similar size, wood strips cut slightly longer than the joist space so that they spring into place (fig. 2-72). They can be spaced 24 to 36 inches apart or as needed. Wire netting nailed between joists may also be used to hold the batt insulation (fig. 2-72). The 1-inch-thick or thicker blanket insulation can also be installed in this way if desired.

Wall Insulation

When blanket or batt insulation for the walls contains a vapor barrier, the barrier should be placed toward the room side and the insulation stapled in place (fig. 2-73A). An additional small

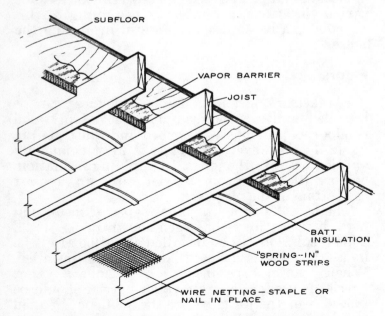

FIGURE 2-72. Installing insulating batts in floor.

strip of vapor barrier at the bottom and top plates and around openings will insure good resistance to vapor movements in these critical areas.

When friction or other insulation without a vapor barrier is used in the walls, the entire interior surface should be covered with a vapor barrier. This is often accomplished by the use of wall-height rolls of 2-or 4-mil plastic film (fig. 2-73B). Other types of vapor barriers extending from bottom plate to top plate are also satisfactory. The studs, the window and door headers, and plates should also be covered with the vapor barrier for full protection. The full-height plastic film is usually carried over the entire window opening and

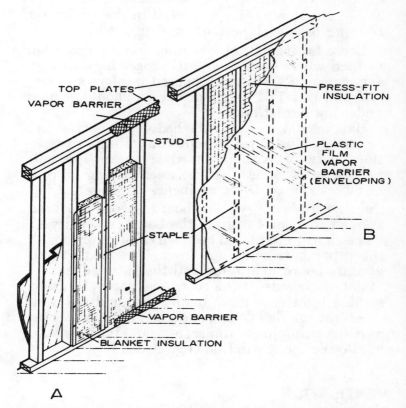

FIGURE 2-73. Wall insulation. A, Blanket insulation with vapor barrier; B, plastic film vapor barrier.

cut out only after the dry wall has been installed. Staple or tack just enough to hold the vapor barrier in place until the interior wall finish is installed.

Ceiling Insulation

Loose-fill or batt-type ceiling insulation is often placed during or after the dry-wall ceiling finish is applied, depending on the roof design. In ceilings having no attic or joist space, such as those with wood roof decking, the insulation is normally a part of the roof construction and may include ceiling tile, thick wood decking, and structural insulating board in various combinations. This type of construction has been covered in the section on framing of low-slope roofs and fig. 2-38.

Loose-fill insulation, blown or hand placed, can be used where there is an attic space high enough for easy access. It can be poured in place and leveled off (fig. 2-74A). A vapor barrier should be used under the insulation.

Batt insulation with attached vapor barrier can be used in most types of roof and ceiling constructions with or without an attic space. The batt insulation is made to fit between ceiling or roof members spaced 16 or 24 inches on center. After the first row of gypsum board sheets has been applied in a level ceiling, the batts (normally 48 inches long) are placed between ceiling joists with the vapor barrier facing down. The next row of gypsum board is applied and the batts added. At the opposite side of the room, the batts should be stapled lightly in place before the final dry-wall sheets are applied. When one set of members serves as both ceiling joists and rafters, an airway should be allowed for ventilation (fig. 2-74B).

VENTILATION

Providing ventilation is an important phase of construction for all houses regardless of their cost.

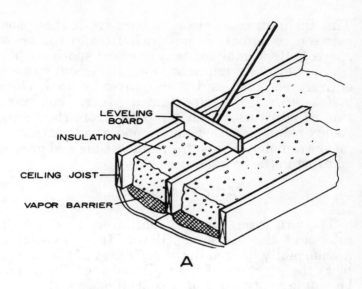
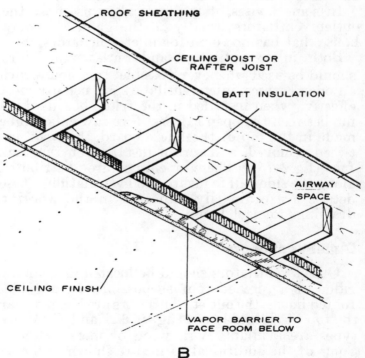

FIGURE 2-74. Ceiling insulation. A, Fill type; B, batt type.

This includes ventilation of attic areas, the spaces between combination joist-rafters, and the crawl space. Only a small amount of crawl-space ventilation is normally required when the crawl space is entirely enclosed and a soil cover is used. Good *attic ventilation* is important to prevent condensation of moisture which can enter from the heated rooms below. Furthermore, good attic ventilation means a cooler attic in the summertime and greater comfort in the living areas below.

Attic Ventilation

The two types of attic ventilators used are the inlet and the outlet ventilator. Inlet ventilators are normally located in the soffit area of the cornice or at the junction of the wall and roof. They may be single ventilators or a continuous *vent*.

In some houses, it is practical to use only the outlet ventilators usually in the gable end of a house that has no room for inlet ventilators.

Both inlet ventilators and outlet ventilators should be used whenever possible. In a house with an open cornice, inlet ventilators are usually most effective when installed in the frieze board which fits between the open rafters. Two saw cuts can be made in the top of the frieze board, the wood between removed, and screen installed on the back face (fig. 2-75). These openings are distributed along the sidewall to insure good ventilation. These details are ordinarily indicated in the working drawings for each house.

Outlet Ventilators

Outlet ventilators should be located as near the ridge of a pitched roof as possible. In a gable-roofed house, the outlet ventilators are located near the top of the gable end (fig. 2-76 A and B.) Many types are available with wood or metal *louvers*. Some of the additional forms are shown in fig. 2-76 C, D, and E. When ladder framing is used on a wide gable overhang, ventilators are often used

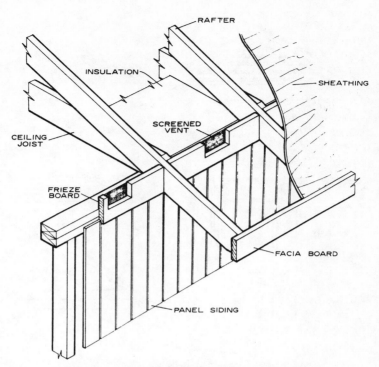

FIGURE 2-75. Inlet ventilators for open cornice.

along the soffit area (fig. 2-76*F*). Ladder framing is formed by lookout members which bear on an interior truss or rafter and on the end wall and extend beyond to a fly rafter. Roof ventilators are available in various forms. They consist of covered and screened metal units designed to fit most roof slopes. They are adapted to gable- or *hip-roof* houses and should be located on the rear slope of the roof as near the ridge as possible.

Inlet Ventilators

Inlet ventilators installed in the soffit area of a closed cornice can consist of individual units spaced as required (fig. 2-77*A*). Another system of inlet ventilation consists of a narrow slot cut in the soffit (fig. 2-77*B*). All ventilators should be screened.

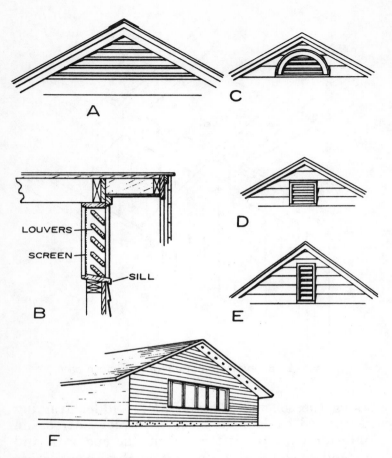

FIGURE 2-76. Outlet ventilators: A, triangular; B, section through louvers; C, half circle; D, square; E, rectangular; F, gable overhang.

Amount of Ventilation

The net or face area of the ventilators required for a house is normally based on some ratio of the ceiling area. The net area is the total area of the ventilator with deductions made for the screening and louvers. When insect screen is used, the total area is reduced by half. In other words, a 1-square-foot ventilator would have a ½-square-foot net area. When louvers are present in addition to the screen, the total net area is only about 40 percent of the total.

The following tabulation shows the total recommended net inlet and outlet areas for various roofs in the form of ratios of total minimum net ventilating area to ceiling area:

Type of roof	Inlet	Outlet
Gable roof with outlet ventilators only		1/300
Gable roof with both inlet and outlet ventilators	1/900	1/900
Hip roof with inlet and outlet ventilators (distributed)	1/900	1/900
Flat or low-slope roofs with ventilators in soffit or eave area only (located at each side of house)		1/250

As an example, assume a house has 900 square feet of ceiling area and a gable roof with a soffit, so both inlet and outlet ventilators can be used. Screen reduces the area by half, so areas required are 1/450 of the ceiling. Thus 900 x 1/450 = 2 square feet. So use two outlet ventilators, each with 1 square foot. The total inlet area (well distributed) should also total 2 square feet.

Crawl-Space Ventilation

As previously mentioned, crawl-space ventilation is not required unless the space is entirely closed. When required, small ventilators can be located in the foundation walls on at least two opposite sides. The amount of ventilation depends on the presence of a ground cover or vapor barrier. When a vapor barrier is used, the total recommended net area of the ventilators is 1/1,500 of the floor area. Use at least two screened vents located on opposite ends of the house. When no vapor barrier is present, the total net area should be 1/150 of the floor area with the ventilators well distributed.

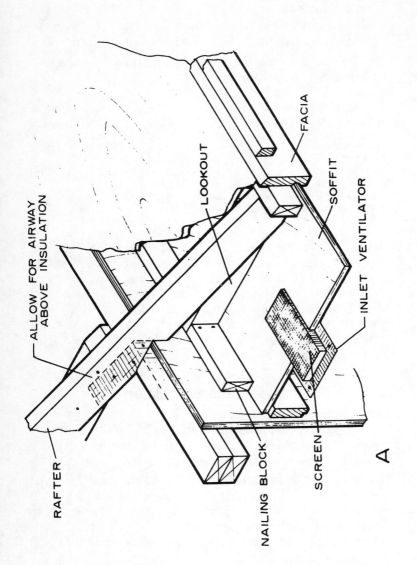

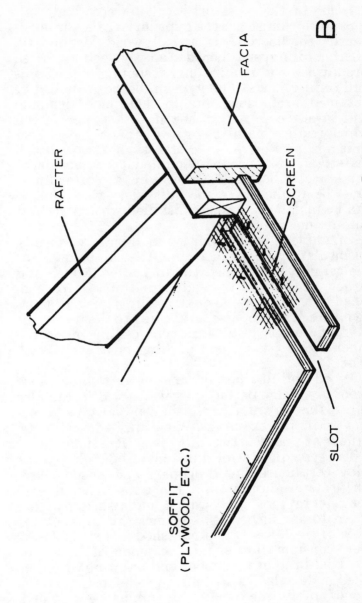

FIGURE 2-77. Soffit inlet ventilators: A, spaced units; B, slot type.

INTERIOR WALL AND CEILING FINISH

Some of the most practical low-cost materials for interior finish are the gypsum boards, the hardboards, the fiberboards, and plywood. Hardboard, fiberboard, gypsum board, and plywood are often prefinished and require only fastening to the studs and ceiling joists. The gypsum boards can also be obtained prefinished, but in their most common and lowest cost form, they have a paper facing which requires painting or wallpapering. A plaster finish is usually more costly than the dry-wall materials and perhaps should not be considered for low-cost houses. Furthermore, a plaster finish must be applied by a specialist. Dry-wall finishes can be installed by the semiskilled workman after a minimum of training.

In addition to the four types of sheet materials, an insulating board or ceiling tile can be used for the ceiling. Prefinished tile in 12- by 12-inch and larger sizes usually requires nailing strips fastened to the underside of the ceiling joists or truss members. Tile can also be applied to the underside of roof boards in a beam-type ceiling or on the inner face of the wood decking on a wood-deck roof.

Wood and fiberboard paneling in tongued-and-grooved V-edge pattern in various widths can also be used as an interior finish, especially as an accent wall, for example. When applied vertically, nailers are used between or over the studs.

The type of interior finish materials selected for a low-cost house should primarily be based on cost. These materials may vary from a low of 5 to 6 cents per square foot for $3/8$-inch unfinished gypsum board to as much as three times this amount for some of the lower cost prefinished materials. However, consideration should be made of the labor involved in joint treatment and painting of unfinished materials. As a result, in some instances, the use of prefinished materials might be justified. These details and material requirements are included in the working drawings or the accompanying specifications for each house. Before interior wall and ceiling finish is applied, insulation should be in place and wiring, heating *ducts*, and other utilities should be roughed in.

Material Requirements

The thickness of interior covering materials depends on the spacing of the studs or joists and the type of material. These requirements are usually a part of the working drawings or the specifications. However, for convenience, the recommended thicknesses for the various materials are listed in the following tabulation based on their use and on the spacing of the fastening members:

Interior Material Finish Thickness

Finish	Minimum material thickness (inches) when framing is spaced	
	16 inches	24 inches
Gypsum board	3/8	1/2
Plywood	1/4	3/8
Hardboard	1/4	
Fiberboard	1/2	3/4
Wood paneling	3/8	1/2

For ceilings, when the long direction of the gypsum board sheet is at right angles to the ceiling joists, use 3/8-inch thickness for 16-inch spacing and 1/2-inch for 24-inch joist spacing. When sheets are parallel, spacing should not exceed 16 inches for 1/2-inch gypsum board. Fiberboard ceiling tile in 1/2-inch thickness requires 12-inch spacing of nailing strips.

Gypsum Board

Application

Gypsum board used for dry-wall finish is a sheet material composed of a gypsum filler faced with paper. Sheets are 4 feet wide and 8 feet long or longer. The edges are usually recessed to accommodate taped joints. The ceiling is usually covered before the wall sheets are applied. Start at one wall and proceed across the room. When batt-type ceiling insulation is used, it can be placed as each row of sheets is applied. Use fivepenny (1⅝-inch-long) cooler-type nails for ½-inch gypsum and fourpenny (1⅜-inch) nails for ⅜-inch gypsum board. Ring-shank nails ⅛ inch shorter than these can also be used. Nailheads should be large enough so that they do not penetrate the surface.

Adjoining sheets should have only a light contact with each other. End joints should be staggered and centered on a ceiling joist or bottom chord of a truss. One or two braces slightly longer than the height of the ceiling can be used to aid in installing the gypsum sheets (fig. 2-78). Nails are spaced 6 to 8 inches apart and should be very lightly dimpled with the hammerhead. Do not break the surface of the paper. Edge or end joints should be double-nailed. Minimum edge nailing distance is ⅜ inch.

Vertical or horizontal application can be used on the walls. Horizontal application of gypsum board is often used when wall-length sheets eliminate vertical joints. The horizontal joint thus requires

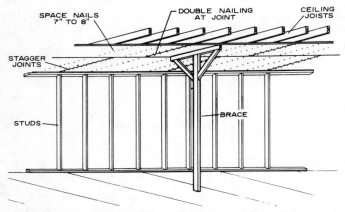

FIGURE 2-78. Installing gypsum board on ceiling.

only taping and treating. For normal application, horizontal joint reinforcing is not required. However, nailing blocks may be used between studs for a damage-resistant joint for the thinner gypsum sheets (fig. 2-79A). Horizontal application is also suitable to the laminated system in which 3/8-inch gypsum sheets are nailed vertically and room-length sheets are applied horizontally with a wallboard or contact adhesive. While this results in an excellent wall, it is much more costly than the single application.

Gypsum board applied vertically should be nailed around the perimeter and at intermediate studs with 1 3/8- or 1 5/8-inch nails, depending on the thickness. Nails should be spaced 6 to 8 inches apart. Joints should be made over the center of a stud with only light contact between adjoining sheets (fig. 2-79B). Another method of fastening the sheets is called the "floating top." In this system, the top horizontal row of nails is eliminated and the top 6 or 8 inches of the sheet are free. This is said to prevent fracture of the gypsum board when there is movement of the framing members.

Joint Treatment

The conventional method of preparing gypsum dry-wall sheets for painting includes the use of a *joint cement* and perforated joint tape. Some gypsum board is supplied with a strip of joint paper along one edge, which is used in place of the tape. After the gypsum board has been installed and each nail driven in a "dimple" fashion (fig. 2-80A), the walls and ceiling are ready for treatment. Joint cement ("spackle" compound), which comes in powder or ready-mixed form, should have a soft putty consistency so that it can be easily spread with a trowel or wide putty knife. Some manufacturers provide a joint cement and a finish joint compound which is more durable and less subject to shrinkage than standard fillers. The procedure for taping (fig. 2-80B) is as follows:

1. Use a wide spackling knife (5-inch) to spread the cement over the tapered and other butt edges, starting at the top of the wall.

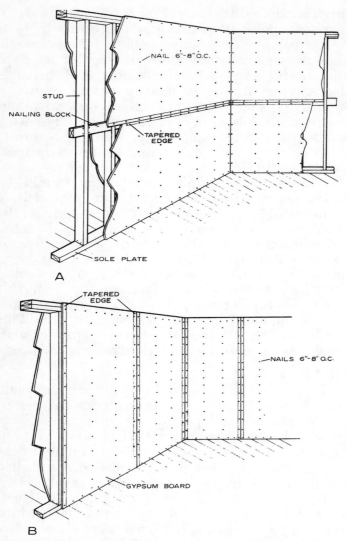

FIGURE 2-79. Installing gypsum board on walls: A, horizontal application; B, vertical application.

2. Press the tape into the recess with the knife until the joint cement is forced through the small perforations.

3. Cover the tape with additional cement to a level surface, feathering the outer edges. When edges are not recessed, apply tape in the normal

manner, but feather out the cement farther so that the joint is level.

4. When dry, sand lightly and apply a thin second coat, feathering the edges again. A third coat may be required after the second has dried.

5. After cement is dry, sand smooth.

6. For hiding nail indentations at members between edges, fill with joint cement. A second coat is usually required.

Interior and exterior corners may be treated with perforated tape. Fold the tape down the center to a right angle (fig. 2-80*C*). Now, (a) apply cement on each side of the corner, (b) press tape in place with the spackle or *putty* knife, and (c) finish with joint cement and sand when dry. Wallboard corner beads of metal or plastic also can be used to serve as a corner guide and provide added strength. They are nailed to outside corners and treated with joint cement. The junction of the wall and ceiling can also be finished with a wood molding in any desired shape, which will eliminate the need for joint treatment (fig. 2-80*D*). Use eightpenny finish nails spaced about 12 to 16 inches apart and nail into the wallplate behind.

Treatment around window and door openings depends on the type of casing used. When a casing bead and trim are used instead of a wood casing, the jambs and the beads may be installed before or during application of the gypsum wall finish. These details will be covered in the section on "Interior Doors, Frames, and Trim."

Plywood and Hardboard

The application of prefinished 4-foot-wide hardboard and plywood sheets is relatively simple. They are normally used vertically and can be fastened with small finish nails (brads). Use nails $1\frac{1}{2}$ inches long for $\frac{1}{4}$- or $\frac{3}{8}$-inch-thick materials and space about 8 to 10 inches apart at all edges and intermediate studs. Edge spacing should be about $\frac{3}{8}$ inch. Set the nails slightly with a nail set.

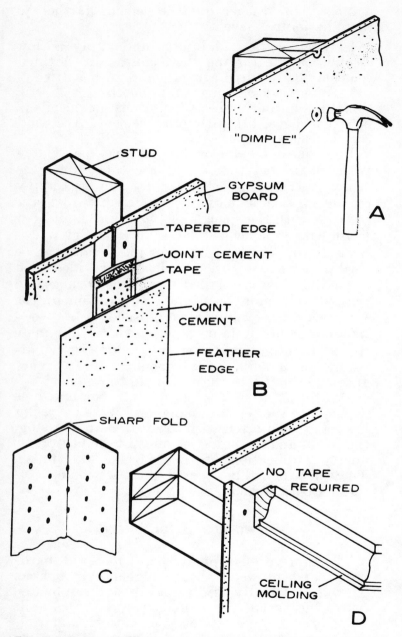

FIGURE 2-80. Preparing gypsum dry-wall sheets for painting. A, Drive nails in "dimple" fashion; B, detail of joint treatment; C, corner tape; D, ceiling molding.

Many prefinished materials are furnished with small nails that require no setting because their heads match the color of the finish.

The use of panel and contact adhesives in applying prefinished sheet materials is becoming more popular and usually eliminates nails, except those used to aline the sheets. Manufacturer's directions should be followed in this method of application.

In applying sheet materials such as hardboard or plywood paneling, it is good practice to insure dry, warm conditions before installing. Furthermore, place the sheets in an upright position against the wall, lined up about as they will be

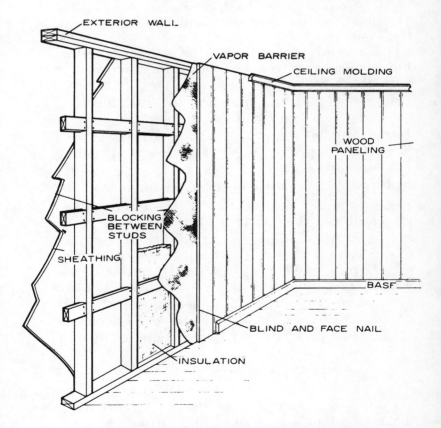

FIGURE 2-81. Application of vertical paneling.

installed, and allow them to take on the condition of the room for at least 24 hours. This is also true for wood or fiberboard paneling covered in the following paragraphs.

Wood or Fiberboard Paneling

Tongued-and-grooved wood or fiberboard (insulating board) paneling in various widths may be applied to walls in full lengths or above the wainscot. Wood paneling should not be too wide (nominal 8-inch) and should be installed at a moisture content of about 8 percent in most areas of the country. However, the moisture content should be about 6 percent in the dry Southwest and about 11 percent in the Southern and Coastal areas of the country. In this type of application, wood strips should be used over the studs or nailing blocks placed between them (fig. 2-81). Space the nailers not more than 24 inches apart.

For wood paneling, use a 1½- to 2-inch finishing or casing nail and blind-nail through the tongue. For nominal 8-inch widths, a face nail may be used near the opposite edge. Fiberboard paneling (planking) is often supplied in 12- and 16-inch widths and is applied in the same manner as the wood paneling. In addition to the blind nail or staple at the tongue, two face nails may be required. These are usually set slightly unless they are color-matched. A 2-inch finish nail is usually satisfactory, depending on the thickness. Panel and contact adhesives may also be used for this type of interior finish, eliminating the majority of the nails except those at the tongue. On outside walls, use a vapor barrier under the paneling (fig. 2-81).

Ceiling Tile

Ceiling tile may be installed in several ways, depending on the type of ceiling or roof construction. When a flat-surfaced backing is present, such as between beams of a beamed ceiling in a low-slope roof, the tiles are fastened with adhesive as recommended by the manufacturer. A small spot of a mastic type of construction adhesive at each corner of a 12- by 12-inch tile is usually sufficient. When tile is edge-matched, stapling is also satisfactory.

A suspended ceiling with small metal or wood hangers, which form supports for 24- by 48-inch or smaller drop-in panels, is another system of applying this type of ceiling finish. It is probably more applicable to the remodeling of older homes with high ceilings, however.

A third, and perhaps the most common, method of installing ceiling tile, is with the use of wood strips nailed across the ceiling joists or roof trusses. These are spaced 12 inches on center. A nominal 1- by 3- or 1- by 4-inch wood member can be used for roof or ceiling members spaced not more than 24 inches on center (fig. 2-82A). A nominal 2- by 2- or 2- by 3-inch member should be satisfactory for truss or ceiling joist spacing of up to 48 inches. Use two sevenpenny or eightpenny nails at each joist for the nominal 1-inch strips and two tenpenny nails for the nominal 2-inch strips. Use a low-density wood, such as the softer pines, as most tile installation is done with staples.

In locating the strips, first measure the width of the room (the distance parallel to the direction of the ceiling joists). If, for example, this is 11 feet 6 inches, use 10 full 12-inch- square tiles and a 9-inch-wide tile at each side edge. Thus, the second wood strips from each side are located so that they center the first row of tiles, which can now be ripped to a width of 9 inches. The last row will also be 9 inches, but do not rip these tiles until the last row is reached so that they fit tightly. The tile can be fitted and arranged the same way for the ends of the room.

Ceiling tiles normally have a tongue on two adjacent sides and a groove on the opposite

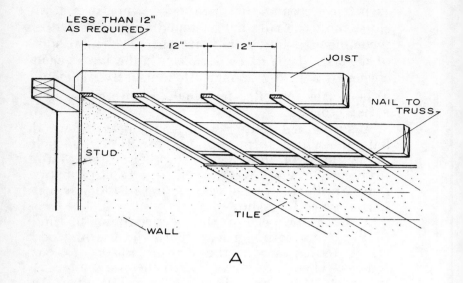

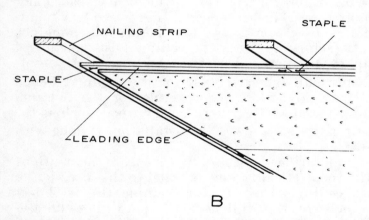

FIGURE 2-82. Ceiling tile installation. A, Nailing strip location; B, stapling.

adjacent sides. Start with the leading edge ahead and to the open side so that they can be stapled to the nailing strips. A small finish nail or adhesive should be used at the edge of the tiles in the first row against the wall. Stapling is done at the leading edge and the side edge of each tile (fig., 2-82B). Use one staple at each wood strip at the leading edge and two at the open side edge. At the opposite wall, a small finish nail or adhesive must again be used to hold the tile in place.

Most ceiling tile of this type has a factory finish, and painting or finishing is not required after it is in place. Because of this, do not soil the surface as it is installed.

Bathroom Wall Covering

When a complete prefabricated shower stall is used in the bathroom instead of a tub, no special wall finish is required. However, if a tub is used, some type of waterproof wall covering is normally required around it to protect the wall. This may consist of several types of finish from a coated hardboard paneling to various ceramic, plastic, and similar tiles.

In the interest of economy, one of the special plastic-surfaced hardboard materials is perhaps a good choice. These are applied in sheet form and fastened with an adhesive ordinarly supplied by the manufacturer of the prefinished board. Plastic or other type moldings are used at the inside corners, at tub edges, at the joints, and as end caps. Several types of calking sealants also provide satisfactory joints.

Other finishes such as ceramic, plastic, and metal tile are installed over a special water-resistant type of gypsum board. Adhesive is spread with a serrated trowel and the $4\frac{1}{4}$- by $4\frac{1}{4}$-inch or other size tile pressed in place. A joint cement is used in the joints of ceramic tile after the adhesive has set. The plastic, metal, or ceramic type of wall covering around the tub area would usually cost somewhat more than plastic-surfaced hardboard. However, almost any type of wall finish can be added at the convenience of the homeowner at any time after the house is constructed.

FLOOR COVERINGS

It is common practice to install wood flooring after the wall finish is applied. This is generally followed by installation of interior trim such as door jambs, casing, base, and other moldings. Wood floors would then be sanded and finished *after* the interior work is completed. Some variation of this is necessary when casing bead and trim are used in dry-wall construction to eliminate the need for standard wood casing around windows and doors. In this instance, the door jambs are installed before the wall finish is applied. Adjustment for the flooring thickness is then made by raising the bottom of the door jambs. Details of the casing bead and trim for dry-wall finishes are included in the next section—"Interior Doors, Frames, and Millwork."

However, for finish floors such as resilient tile and prefinished wood block flooring, it is usually necessary to have most, if not all, the interior work completed before installation. This is especially true of resilient flooring. Manufacturers' recommendations usually state that all tradesmen will have completed their work, even the painters, before the resilient tile is installed. In such cases, a resilient cove base might be a part of the finish, rather than the conventional wood base.

Types of Flooring

The term "finish flooring" usually applies to the material used as the final wearing surface. In its simplest form, it may be merely a *sealer* or a paint finish applied to a tongued-and-grooved plywood subfloor. Perhaps one of the most practical floor coverings for low-cost houses is some type of resilient tile such as asphalt, rubber, vinyl asbestos, etc. These materials may vary a great deal in cost. Because they are applied with an adhesive, the installation costs are usually quite low when compared to other finish flooring. However, when prices are competitive with other materials, wood finish floors in the various patterns might be selected.

Hardwood strip flooring in the best grades is used in many higher cost houses (fig. 2-83*A*). Thinner tongued-and-grooved strip flooring (fig. 2-83*B*) and thin square-edged flooring (fig. 2-83C) are lower in cost than $^{25}/_{32}$-inch strip flooring and might be considered for use initially or at a later date. The use of low-grade softwood strip flooring with a natural or painted finish can also be considered. However, the installation costs of these materials

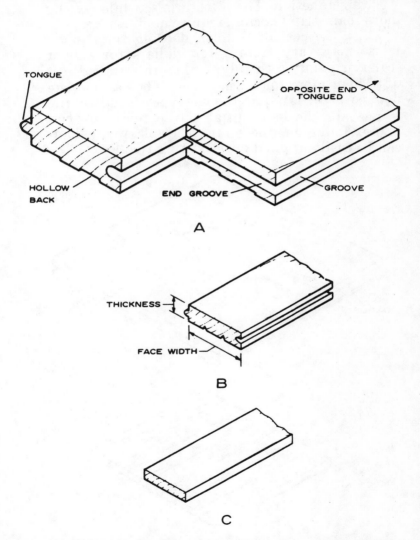

FIGURE 2-83. Strip flooring. A, Side and end matched; B, side matched; C, square edged.

are usually higher than those for resilient coverings. Many types require sanding and finishing after they are nailed in place. Another wood flooring, the parquet or wood block floor (fig. 2-84), is usually prefinished, is supplied in 6- by 6- or 9- by 9-inch squares, and is installed by nailing or with an adhesive. These materials, while costing more than strip flooring, require no finishing.

Thus, the cost of a finish flooring for a small house may vary from a few dollars for a floor sealer or paint on the plywood to an installed cost of $350 or more for finish floors. When a complete floor covering is required, some type of resilient tile is perhaps the best initial choice for a low-cost house. Later on, the more desirable wood floors can be easily applied over existing tile floors.

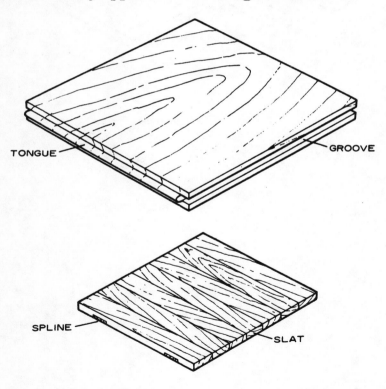

FIGURE 2-84. Wood block flooring.

INSTALLATION OF FLOORING

Strip Flooring

Before laying strip flooring, be sure that the subfloor is clean and covered with a *building paper* when a board subfloor is used. This will aid in preventing air infiltration and help maintain a comfortable temperature at the floor level when a crawl space is used. The building paper should be chalklined at the joists as a guide in nailing the strip flooring. Before laying the flooring, the bundles should be opened and the flooring spread out and exposed to a warm, dry condition for at least 24 hours, and preferably 48 hours. Moisture content of the flooring should be 6 percent for the dry Southwest area, 10 percent for Southern and Coastal areas, and 7 percent for the remainder of the country.

Strip flooring should be laid at right angles to the joists, (fig. 2-85A). Start at one wall, placing the fiirst board $1/2$ to $5/8$ inch away from the wall and face-nail so that the base and shoe will cover the space and nails (fig. 2-85B). Next, blind-nail the first strip of flooring and each subsequent piece with eightpenny flooring nails (for $25/32$-inch thickness) at each joist crossing. Nail through the tongue and into the joist below with a nail angle of 45° to 50°. Set each nail to the surface of the tongue. Each piece should be driven up lightly for full contact. Use a hammer and a short piece of scrap flooring to protect the edge. The last flooring strip must be face-nailed at the wall line, again allowing $1/2$- to $5/8$-inch space for expansion of the flooring.

Other thinner types of strip flooring are nailed in the same way, except that sixpenny flooring nails may be sufficient. Square-edge flooring must be face-nailed using two $1 1/2$-inch finish nails (brads) on about 12-inch centers.

Wood Block Flooring

Wood block flooring with matching tongued-and-grooved edges is often fastened by nailing, but

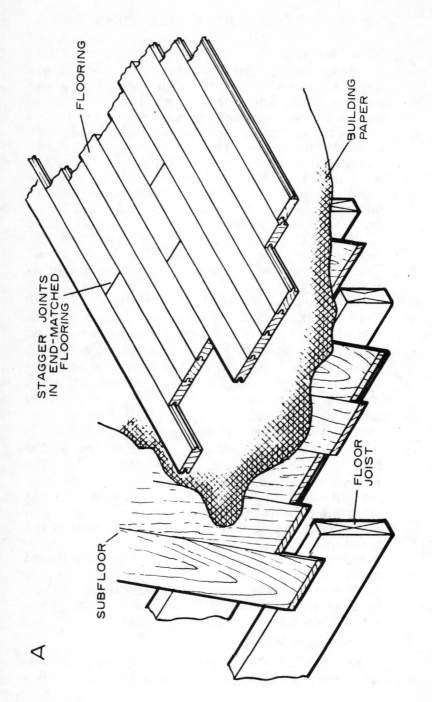

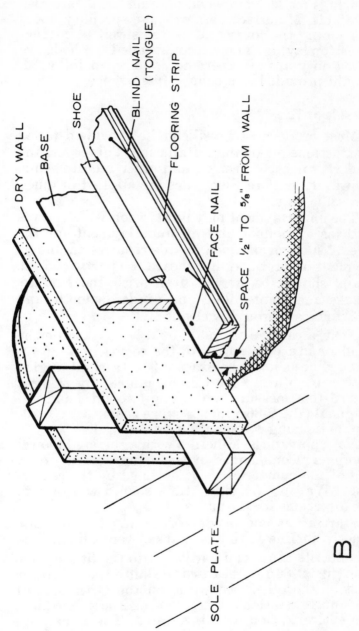

FIGURE 2-85. Installing strip flooring: A, general application; B, laying first strip.

may be installed with an adhesive. Other block floorings made from wood-base materials are also available. Manufacturers of these specialty floors can supply the correct adhesive as well as instructions for laying. Their recommended methods are based on years of experience, and when followed, should provide maintenance-free service.

Resilient Tile

Most producers of resilient tile provide detailed instructions regarding installation. This covers the underlayment, adhesives, and other requirements. However, most resilient tiles are applied in much the same way.

The underlayment may be plywood which serves both as a subfloor and an underlayment for the tile. A subfloor of wood boards requires an underlayment of plywood, hardboard, or particleboard. Nails should be driven flush with the surface, cracks and joints filled and sanded smooth, and the surface thoroughly cleaned.

Next, a center baseline should be marked on the subfloor in each direction of the room (fig. 2-86A). Centerlines should be exactly 90° (a right angle) with each other. This can be assured by using a 3:4:5-foot measurement along two sides and the diagonal (fig. 2-86A). In a large room, a 6:8:10-foot measuring combination can be used.

Now, spread the adhesive with a serrated trowel (both as recommended by the manufacturer) over one of the quarter-sections outlined by the centerlines. Waiting (drying) time should conform to the directions for the adhesive.

Starting at one inside corner, lay the first tile exactly in line with the marked centerlines. The second tile can be laid adjacent to the first on one side (fig. 2-86B). The third tile laid adjacent to the first at the other centerline on the other side of the quarter section. Thus, in checkerboard fashion, the entire section can be covered. The remaining three sections can be covered in the same way.

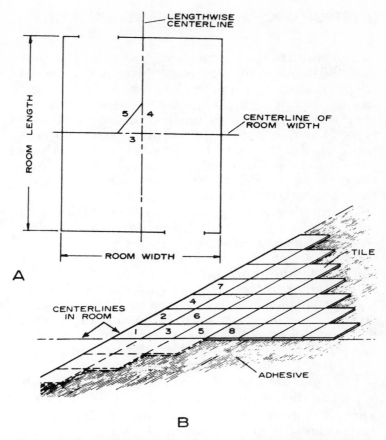

FIGURE 2-86. Installing resilient tile: A, center baseline; B, order of laying tile.

Some tile requires only pressing in place; others should be rolled after installing for better adherence.

The edge tiles around the perimeter of the room must be trimmed to fit to the edge of the wall. A clearance of ⅛ to ¼ inch should be allowed at all sides for expansion. This edge is covered with a cove base of the same resilient material or with a standard wood base. Wood base is usually lower in cost than the resilient cove base, but installation costs are somewhat greater.

INTERIOR DOORS, FRAMES, AND MILLWORK

The cost of interior finish such as cabinets, door frames, doors, and moldings varies a great deal with wood species and styles. For example, pine casing used as trim for door and window frames may cost only half as much as some hardwood trim. However, some types of interior panel doors made of pine may cost twice as much as a mahogany flush door. Choice of materials should be based both on cost and utility. Such details are covered in the building plans and specifications.

The number and types of cabinets used in a house can make a substantial difference in the overall cost. Prefinished base cabinets with counter and wall cabinets for kitchens may range in price from $30 to $35 and more per lineal foot. This should be compared to the lower cost of simple shelving with provisions for installation of the doors at a later date; this may mean a saving of several hundred dollars. When a cost reduction is needed to keep within the limits of available funds, some such substitution may be necessary.

The moisture content of all interior wood finish when installed is important in the overall performance. The recommended moisture content for interior finish varies between 6 and 11 percent, depending on climatic conditions. Recommended moisture content is the same as outlined for wood and fiberboard paneling in the section on "Interior Wall and Ceiling Finish."

Interior Doors and Trim

Door frames.—The rough door openings provided for during framing of the interior walls should accommodate the assembled frames. The allowance was $2\frac{1}{2}$ inches plus the door width and 3 inches plus the door height. When thin resilient tile is used over the subfloor, the allowance is $2\frac{1}{4}$ inches for door height. Frames consist of a head *jamb* with two side jambs and the stops. When a wood casing is used around the door frame as trim, the width of the jambs is the same as the overall wall thickness. When a metal casing is used with the dry wall, eliminating the need for the wood casing, the jamb width is the same as the stud depth width. The side and head jambs and the stop are assembled as shown in fig. 2-87A. Jambs may be purchased in sets or can be easily made in a small shop with a table or radial-arm saw.

Casing.—Casing is the trim around the door opening. It is nailed to the edge of the jamb and to the door buck (edge stud). A number of shapes are available, such as colonial (fig. 2-87B), ranch (fig. 2-87C), and plain (fig. 2-87D). Casing widths vary from $2\frac{1}{4}$ to $3\frac{1}{2}$ inches, depending on the style. Thicknesses vary from $\frac{1}{2}$ to $\frac{3}{4}$ inch. A casing bead or metal casing used to trim the edges of the gypsum board at the door and window jambs (fig. 2-87A) eliminates the need for the wood casing.

Doors.—Two general styles of interior doors are the panel and the flush door. The interior flush door is normally the hollow-core type (fig. 2-88A), which costs much less than the solid core. The five-cross-panel door (fig. 2-88B) is usually lower in cost than the colonial type (fig. 2-88C).

Standard widths for interior doors are: (a) Bedroom and other rooms, 2 feet 6 inches; (b) bathroom, 2 feet 4 inches; (c) small closet and linen closet, 2 feet. Interior door heights should be the same as for exterior doors; standard height is 6 feet 8 inches. Doors of these sizes can be obtained from most lumber dealers.

Doors should be hinged so that they open in the direction of natural entry. They should also swing against a blank wall whenever possible, never into a hallway. Door swing directions and sizes are shown on the working drawings.

Installation of Doors

Frames.—When jambs are not preassembled, side jambs are nailed to the head jamb with three

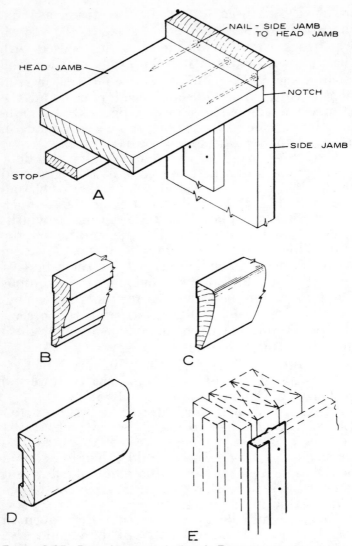

FIGURE 2-87. Door frame and trim. A, Frame components and assembly; B, colonial casing; V, ranch casing; D, plain casing; E, metal casing.

eightpenny coated nails (fig. 2-87A). Cut the side jambs to the correct length before nailing. A temporary brace can then be nailed across the bottom of the side jambs so the width is the same throughout the full height of the door.

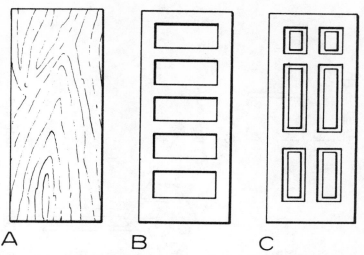

FIGURE 2-88. Interior doors: A, flush; B, five-cross-panel; C, colonial, panel type.

The frame is now placed in the opening and fastened to the wall studs with the aid of shingle wedges used between the side jamb and the rough door buck (side stud) (fig. 2-89A). Plumb and fasten one side first, using four or five sets of wedges along the height, and nail the jamb with pairs of eightpenny finish nails at each wedged area. Square the top corners of the opening and fasten the opposite jamb in the same manner. Use a straightedge along the face of the jambs when lining them up with the wedges.

Casings.—Casings are nailed to the edges of the jamb and to the door buck. First, cut off the shingle wedges flush with the wall. Position the casing with about a $3/16$-inch edge distance from the face of the jamb (fig. 2-89A). Depending on the thickness of the casing, sixpenny or sevenpenny casing or finish nails should be used. When the casing has one thin edge, use a $1\frac{1}{2}$-inch brad or finish nail along this edge. Space the nails in pairs about 16 inches apart.

Casings with molded forms (fig. 2-87 B and C) must have a *mitered joint* where head and side cas-

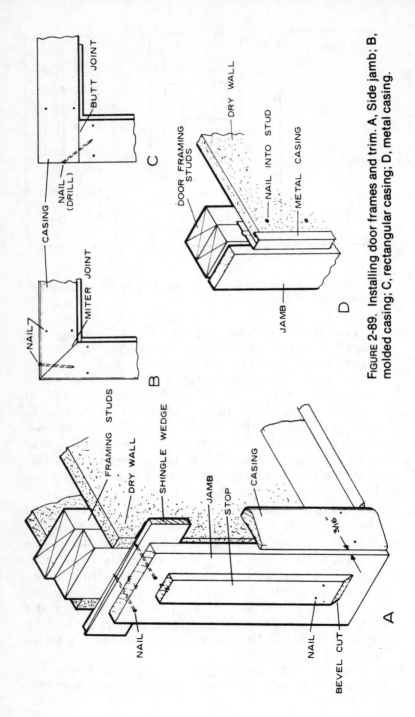

FIGURE 2-89. Installing door frames and trim. A, Side jamb; B, molded casing; C, rectangular casing; D, metal casing.

ings join (fig. 2-89*B*). Rectangular casing can be butt-joined (fig. 2-89*C*). Casing for the interior side of exterior door frames is installed in the same way.

Metal casing used instead of wood casing can be nailed in place in two ways. After the jambs have been installed and before the gypsum board is placed, the metal casing can be nailed to the door buck around the opening (fig. 2-87*E*). The gypsum board is then inserted into the groove and nailed to the studs in normal fashion. The second method consists of placing the metal casing over the edge of the gypsum board, positioning the sheet properly with respect to the jamb and nailing through both the gypsum board and the casing into the stud behind (fig. 2-89*D*). Use the same type nails and nail spacing as described in the section on "Interior Wall and Ceiling Finish" for gypsum board.

Doors and stops.—The door opening is now complete. Without the door in place, it is often referred to as a "cased opening." When cost is a major factor, the use of this opening with a curtain rod and curtain will insure some privacy. Door stops can be fitted and doors hung later.

The door is fitted between the jambs by planing the sides to the correct width. This requires a careful measurement of the width before planing. When the correct width is obtained, the top is squared and trimmed to fit the opening. The bottom of the door is now sawed off with the proper floor clearance. The side, top, and bottom clearances and the location of the hinges are shown in figure 90. Remember that paint or finish takes up some of the space, so allow for this.

The narrow wood strips used as stops for the door are usually $7/16$ inch thick and may be $1\frac{1}{2}$ to $2\frac{1}{4}$ inches wide. They are installed on the jambs with a mitered joint at the junction of the head and side jambs. A 45° bevel cut at the bottom of stops 1 to $1\frac{1}{2}$ inches above the finish floor will eliminate a dirt pocket and make cleaning or refinishing of the floor much easier (fig. 2-89*A*).

Before fitting the exterior doors, install a threshold between side jambs to cover the junction of the sill and the flooring (or allow for the threshold). Nail to the floor and sill with finish nails.

Installation of Door Hardware

Four types of hardware sets available in a number of finishes are commonly used for doors. They are classed as: (a) Entry lock for exterior doors, (b) bathroom set (inside lock control with a safety slot for opening from the outside), (c) bedroom lock set (keyed lock), and (d) passage set (without lock). Two hinges are normally used for 1⅜-inch interior doors. To minimize warping of exterior doors in cold climates, use three hinges. Exterior doors, 1¾ inches thick, require 4- by 4-inch loose-pin hinges and the 1⅜-inch interior doors 3½- by 3½-inch loose-pin hinges.

Hinges.—Hinges are *routed* or *mortised* into the edge of the door with about a 3/16- or ¼-inch back spacing (fig. 2-91A). This may vary slightly, however. Adjustments should be made, if necessary, to provide sufficient edge distance so that screws have good penetration in the wood. Locate the hinges as shown in figure 90 and use one hinge half to mark the outline of the cut. If a router is not available, mark the hinge outline and the depth of the cut and remove the wood with a wood chisel. The depth of the routing should be such that the surface of the hinge is flush with the wood surface. Screws are included with each hinge set and should be used to fasten the hinge halves in place.

The door is now placed in the opening and blocked for the proper clearances. Stops should be tacked in place temporarily so that the door surface is flush with the edge of the jamb. Mark the location of the door hinges on the jambs, remove the door, and with the remaining hinge halves and a small square mark the outline. The depth of the routed area should be the same as that on the door (the thickness of the hinge half). After fastening the hinge halves, place the door in the

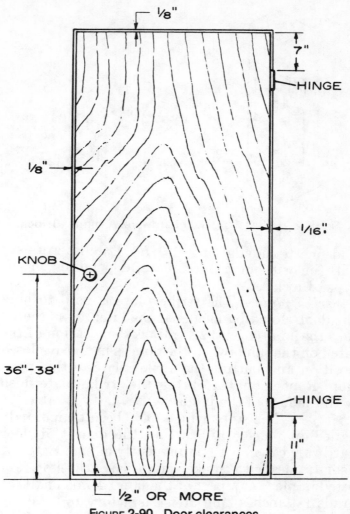

FIGURE 2-90. Door clearances.

opening and insert the pins. If marking and routing were done correctly, the door should fit perfectly and swing freely.

Locks.—Door lock and latch sets are supplied with paper templates which provide the exact location of holes for the lock and latch. Follow the printed directions, locating the door knob 36 to 38 inches above the floor (fig. 2-90). Most lock sets require only one hole through the face of the door

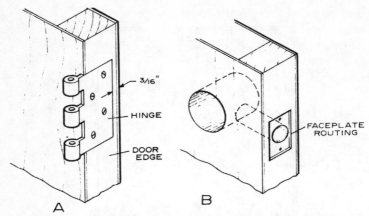

FIGURE 2-91. Installing door hardware: A, hinge; B, lock.

and one at the edge (fig. 2-91*B*). Latches are used with or without a face plate, depending on the type of lock set.

Strike plate.—The strike plate is used to hold the door closed by means of the latch. It is routed into the jamb (fig. 2-92*A*). Mark the location of the latch on the jamb when the door is in a near-closed position and outline the strike plate to this position. Rout enough to bring the strike plate flush with the face of the jamb.

Stops.—The stops which have been temporarily nailed in place while fitting the door and door hardware can now be permanently nailed. Exterior door frames with thicker jambs have the stop rabbeted in place as a part of the jamb. Finish nails or brads 1½ inches long are satisfactory for nailing. The stop at the lock side should be nailed first, setting it against the door face when the door is latched. Use nails in pairs spaced about 16 inches apart. The clearances and stop locations shown in fig. 2-92*B* should be generally followed.

Window Trim

The casing used around the window frames on the interior of the house is usually the same pattern as that used for the interior door frames.

There are two common methods of installing wood trim at window areas: (a) With a *stool* and *apron* (fig. 2-93A) and (b) with complete casing trim (fig. 2-93B). Metal casing is also used around the entire window opening.

In a prefitted double-hung window, the stool is normally the first piece of trim to be installed. It is notched out between the jambs so that the forward edge contacts the lower sach rail (fig. 2-93A). Of course, in windows that are not preassembled, the sash must be fitted before the trim is installed. The stool is now blind-nailed at the ends with eightpenny finish nails so that the casing at the side will cover the nailheads. With hardwood, predrilling is usually required to prevent splitting. The stool should also be nailed at midpoint to the sill, and later to the apron when it is installed. Toenailing may be substituted for face-nailing to the sill (fig. 2-93A).

The casing is now applied and nailed as described for the door frames, except that the inner edge is flush with the inner face of the jambs so that the *stop* covers the joint (fig. 2-93A). The stops are now fitted similarly to the interior door stops and placed against the lower sash so that it can slide freely. A 1½-inch casing nail or brad should be used. Use nails in pairs spaced about 12 inches apart. When full-length weatherstrips are included with the window unit, locate the stops against them to provide a small amount of pressure. Cut the apron to a length equal to the outer width of the casing line (fig. 2-93A) and nail to the framing sill below with eightpenny finish nails.

When casing is used to finish the bottom of the window frame instead of the stool and apron, a narrow stool or stop is substituted (fig. 2-93B). Casing along the bottom of the window is then applied in the same way as at the side and head of the window.

When metal casing is used as trim around window openings, it is applied to the sill as well as at the side and head of the frame. Consequently, the jambs and sill of the frame are not as deep as

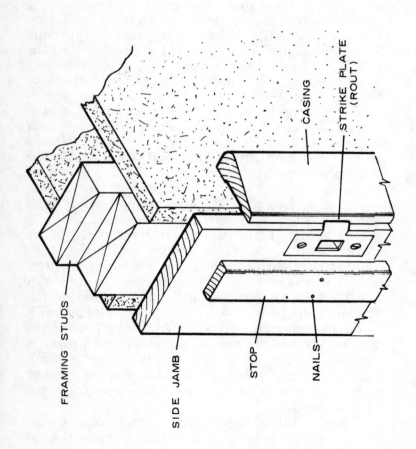

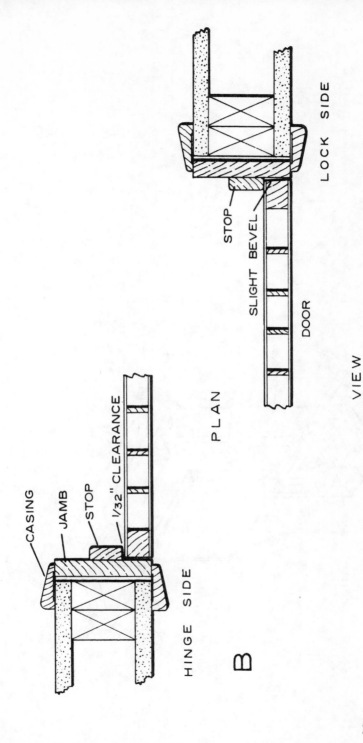

Figure 2-92. Door installation. A, Stike plate; B, stop clearances.

295

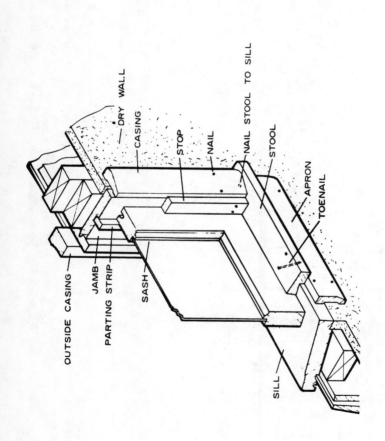

A

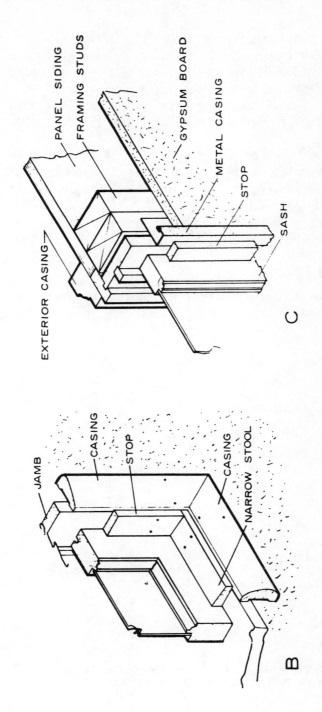

FIGURE 2-93. Window trim. A, With stool and apron; B, wood casing at window; C, metal casing at windows.

when wood casing is used. The stops are also narrower by the thickness of the dry-wall finish. The metal casing is used flush with the inside edge of the window jamb (fig. 2-93C). This type of trim is installed at the same time as the dry wall and in the same way described for the interior door frames.

Other types of windows, such as the awning or hopper types or the casement, are trimmed about the same as the double-hung window. Casing of the same types shown in fig. 2-87 B, C, D, and E can also be used for these units.

Base Moldings

Types

Some type of trim or finish is normally used at the junction of the wall and the floor. This can

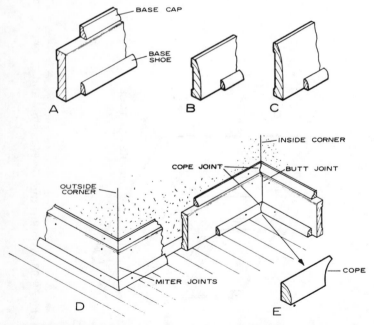

FIGURE 2-94. Base moldings. A, Two-piece; B, narrow; C, medium width; C, installation; E, coped joint.

consist of a simple wood member which serves as both the *base* and *base shoe* or a more elaborate two-piece unit with a square-edge base and base cap (fig. 2-94A). One-piece standard base may be obtained in a 2¼-inch width (fig. 2-94B) or 3¼-inch medium width (fig. 2-94C). The true base shoe, sometimes called *quarter-round*, is not actually quarter-round as it is ½ by ¾ inch in size. In the interest of economy, the base shoe can be eliminated, using only a single-piece base. Resilient floors may be finished with a simple narrow wood base or with a resilient *cove* base which is installed with adhesives.

Installation

Wide square-edge *baseboard* should be installed with a butt joint at inside corners and mitered joint at outside corners (fig. 2-94D). It should be nailed to each stud with two eightpenny finishing nails. Molded single-piece base, *base molding*, and *base shoe* should have a coped joint at inside corners and a mitered joint at outside corners. A *coped joint* is one in which the first piece has a square-cut end against the wall and the second member at the inside corner has a coped joint. This is accomplished by sawing a 45° miter cut and, with a coping saw, trimming along the mitered edge to fit the adjoining molding (fig. 2-94E). This results in a tight inside joint. The base shoe should be nailed to the subfloor with long slender nails, and not to the baseboard itself. This will prevent openings between the shoe and the floor if floor joists dry out and shrink.

Cabinets

As discussed in the first part of this section, kitchen and other cabinets can increase the cost of a house substantially. Thus, in many cases, it may be necessary to use the most simple forms of storage areas to remain within a limited budget. However, even when the cabinets must be quite simple,

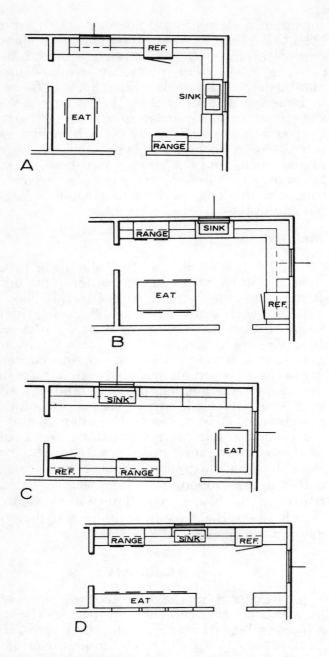

FIGURE 2-95. Kitchen arrangements. A, U-type; B, L-type; parallel wall; D, sidewall.

it is good practice to provide for additions which can be made later such as the installation of the doors and door hardware.

Kitchen Arrangements

A kitchen, no matter how small, should be laid out so that there is a good relationship between the refrigerator, the sink, and the stove. This is desirable in order to save steps and time in the preparation of meals. Kitchen sizes and shapes vary a great deal and usually control the arrangement of the utilities and the cabinets. Fig. 2-95 shows the four common types of kitchen layouts: (a) U-type, (b) L-type, (c) parallel wall, and (d) sidewall type. The L-type and the sidewall type (fig. 2-95 B or D) are perhaps those most adaptable to a small kitchen. Kitchen layouts are shown on the working drawings for each house.

Kitchen Cabinet Units

Kitchen cabinets are normally classed as (a) base cabinets and (b) wall cabinets. The base or floor unit, which contains a counter, may include both drawers and doors and shelves in various combinations. The wall units are normally a series of shelves with hinged or sliding doors located above the base cabinets. The proportions and sizes of these cabinets are shown in fig. 2-96. A standard height for the base unit is 36 inches, with a top width of 25 inches. A low-cost counter top may consist simply of plywood with a resilient surface covering. The more elaborate tops usually have a plastic laminated surface with a molded edge and backsplash.

Construction of Low-Cost Kitchen Cabinets

Low-cost kitchen cabinets can consist of a series of shelves enclosed with vertical dividers and ends.

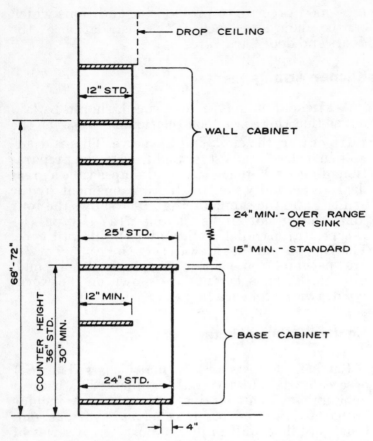

FIGURE 2-96. Kitchen cabinet proportions.

They can be designed so that flush doors can be added later at little cost and with no alterations. Curtains across the top of the opening can be temporarily used.

Base units.—Fig. 2-97 shows the details of a simple base unit. Any combination of opening widths and shelf spacing can be used. Ripping 4-foot-wide sheets of plywood or particleboard in half provides material for ends, dividers, and shelves. Use ⅝ or ¾-inch thickness in an AC grade when only one side is exposed. When shelf layout is decided, saw slots across the width of the ends and dividers so that shelves will fit them. These *dados* are usually ¼ inch deep. Provide for about a 3½- by 3½-inch toe space at the bottom. When assembling, nail the ends and dividers to the shelves with eightpenny nails spaced 5 to 6 inches apart. Use finish nails where ends are exposed. A 1- by 3-inch rail is used at the top in front to aid in fastening the top to the cabinet (fig. 2-97*A*). This will later serve to frame the doors. A 1- by 4-inch cleat is used across the back at the top, serving to fasten the cabinet to the wall studs, as well as for fastening the top (fig. 2-97*B*). The ends and dividers are notched to receive this cleat (fig. 2-97*C*). On finished ends, this notch is cut only halfway through. Fasten the cabinet to the wall by nailing or screwing through the back cleat into each stud with eightpenny nails or 2-inch screws.

The top of plywood or particleboard is nailed in place onto the ends, back cleat, and top rail with eightpenny finish nails. When the top has a plastic finish, use small metal angles around the interior and fasten them to the cabinet and top with small screws. When side stiles and doors are added later, a pleasant, utilitarian cabinet will result (fig. 2-97*D*). Until then, however, curtains can be used across the openings.

Any combination of base cabinets can be constructed. The sink cabinet usually consists only of a pair of doors and a bottom shelf. Vent slots are usually cut in the top cross rail. A cutout is necessary in the top to provide for a self-rimming or rimmed sink.

Wall units.—Wall units are about 30 inches high when installed below a drop ceiling. When carried to the ceiling line, they may be 44 or 45 inches in height. The depth is usually about 12 inches. A 4-foot-wide sheet of plywood or particleboard can be ripped in four pieces if this depth is used. As in base unit construction, the wall cabinet can consist of dadoed ends and one or more vertical dividers nailed to the shelves (fig. 2-98*A*). Use eightpenny finish nails when the end is exposed. The cleat is used to fasten the cabinet to the wall (fig. 2-98*B*).

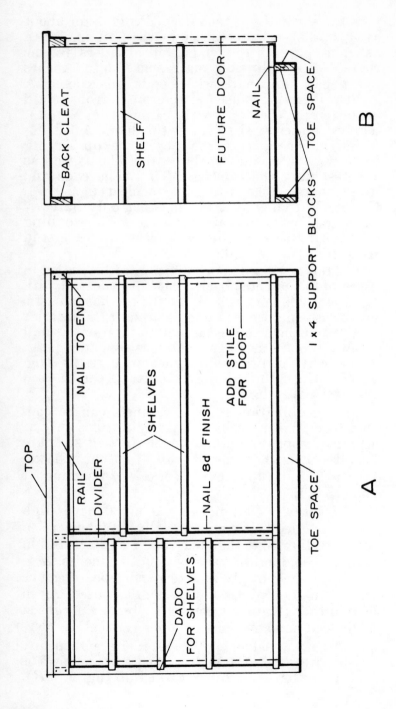

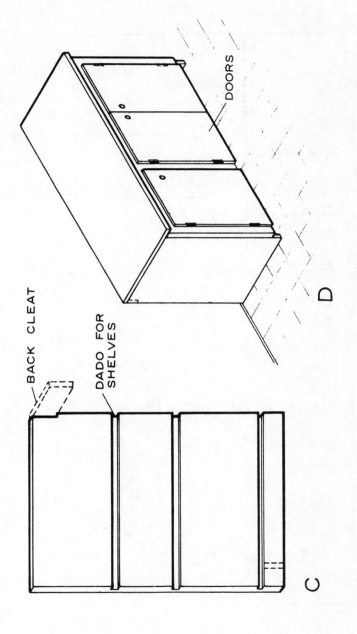

FIGURE 2-97. Kitchen cabinet base unit. A, Front view; B, section; C, end from interior; D, overall view.

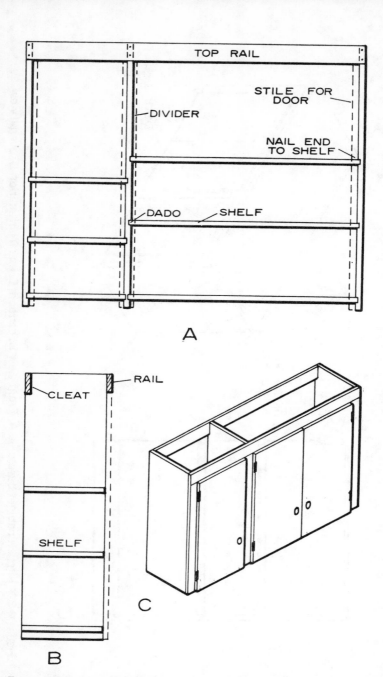

FIGURE 2-98. Kitchen cabinet wall unit. A, Front view; B, section; C, overall view.

A top rail ties the sides and dividers together and, when the vertical stiles are added, provides framing for doors (fig. 2-98*C*).

Wardrobe-Closet Combinations

Closets are often eliminated in a low-cost house for economic reasons. A closet requires wall framing, interior and exterior covering, a door frame and door, and trim. However, some type of storage area should be included which will be pleasant in appearance, yet low in cost. Practical storage areas, often called wardrobe-closets, consist of wood or plywood sides, shelves, and some type of

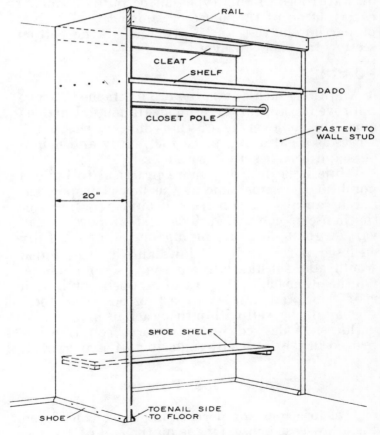

FIGURE 2-99. Wardrobe closet.

wood or plywood sides, shelves, and some type of door or curtain. Figure 2-99 shows a simple built-in wardrobe that can be initially curtained. A folding door unit can be added later.

Wardrobe-closets are normally built of 5/8- or 3/4-inch plywood or particleboard. Provide dados for the shelf or shelves, as used on the kitchen cabinets. A back cleat at the top provides a member for fastening the unit to the wall. The sides can be trimmed at the floor with a base shoe molding which, with toenailing, keeps the sides in place. Add a simple closet pole. Fasten the unit to the wall at the top cleat and, when in a corner, also to the side wall (fig. 2-99). Any size or combination of wardrobes of this type can be built in during construction of the house or added later. Shelves or partial shelves can be provided in the bottom section for shoe storage.

PORCHES

Additions such as a *porch* or an attached garage improve the appearance of a small house. Furthermore, in some areas of the country, the porch serves as a gathering place for family and neighbors and becomes almost a necessity.

While it is probably more practical to build a porch at the same time as the house proper, the porch can also be constructed quite readily after the house is completed. The cost of porches can vary a great deal, depending on design. A fully enclosed porch with windows and interior finish would add substantially to the cost of a house. On the other hand, the cost of an open porch with roof, floor, and simple supporting posts and footings might be well within the reach of many home builders. If desired, an open porch could be improved in the future by enclosing all or part of it.

Types of Open Porches

Two locations for porches might be considered for a low-cost house. One is on the end of a house (fig. 2-100A). Another is at the side of a house

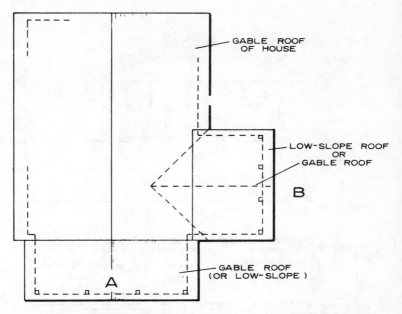

FIGURE 2-100. Porch locations: A, end of house; B, side of house.

(fig. 2-100*B*). The porch at the end of a gable-roofed house may have a gable roof similar to the house itself. The slope of the roof should usually be the same as the main roof. When the size of the porch is somewhat less than the width of the house, the roofline will be slightly lower (fig. 2-101). When the porch is the same width, the roofline of the house is usually carried over the porch itself. A low-pitch roof in a shed or hip style can also be used for a porch located at the end of a house.

A low-slope roof can also be used for a porch at the side of a gable-roofed house (fig. 2-102). This system of room construction is usually less expensive than other methods, as the rafters also serve as ceiling joists.

Another method of roofing a porch at the side of the house is with the gable roof (fig. 2-103). While somewhat more costly than a flat or low-pitch roof, it is probably more pleasing in appearance. Both rafters and ceiling joists are normally used in this

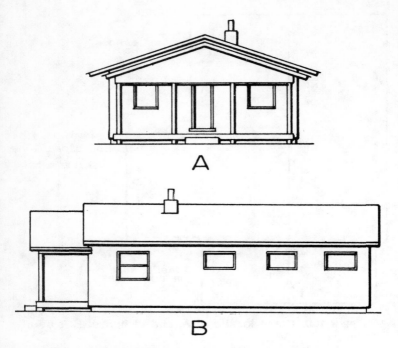

FIGURE 2-101. End porch. A, End view; B, side view.

system. Thus, if a porch is to be a part of the original house or constructed later, select the style most suited to the design of the house and to the available funds.

Construction of Porches

While many types of foundations might be used for a porch as well as for the house, we will consider only the post or pier type because of its lower cost. A full-masonry foundation wall with proper footings will increase the cost substantially over the post or pier type. With a proper soil cover and a skirtboard of some type, the performance and appearance of such a post or pier foundation will not vary greatly from that of a full-masonry wall.

Floor System

The floor system for the open porch should consist of treated wood posts or masonry piers con-

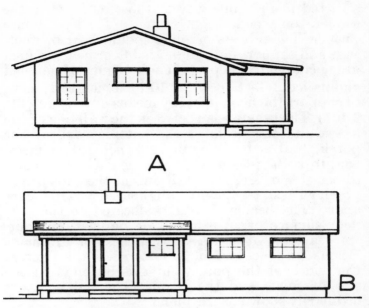

FIGURE 2-102. Side porch with low-scope roof. A, End view; B, side view.

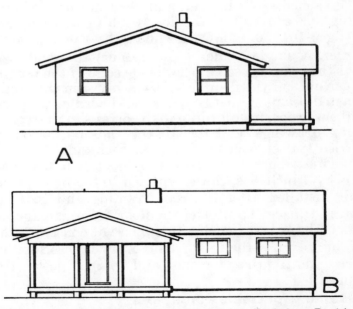

FIGURE 2-103. Side porch with gable roof. A, End view; B, side view.

structed in the same way as those outlined in the section on foundation systems for the house proper. These posts or piers should bear on footings and support nail-laminated floor beams which are anchored to the posts. These beams, of doubled and nailed 2- by 8- or 2- by 10-inch members, serve to support the floor joists by means of *ledgers* (fig. 2-104). The beams span between the wall line of the house and the posts at the outside edge of the porch. Nail 2- by 3- or 2- by 4-inch ledger members to both sides of the intermediate beams and to one side of single edge beams at the ends of the porch with sixteenpenny nails spaced 8 to 10 inches apart. The beams should be located so that the floor surface of an 8-foot-wide porch slopes outward a total of at least 1½ inches for drainage. Beams are spaced about 10 feet or less apart across the width of the porch, but this depends on the size of joists and the beams. These details are ordinarily shown on the plans.

Joist and beam tables can also be used to determine the correct span-spacing-size relationship. Use a metal joist hanger or angle iron when fastening the beam ends to the floor framing of the house (fig. 2-104), or allow the ends to bear on a post or pier. A single header (with ledger) the same size as the beam is used at the ouside edge of the porch.

Now, cut the floor joists to fit between the laminated beams so that they rest on the ledgers. Space them properly according to the details in the working drawings, and toenail the ends to the beam with two eightpenny nails on each side.

Dressed and matched porch flooring, in nominal 1- by 4-inch size, can be applied to the floor joists or installed after the roof framing and roofing are in place. To protect the floor from damage, it is perhaps best to delay this phase of construction. This requires the use of temporary braces for the roof until the flooring and porch posts are installed.

Framing for Gable Porch Roof

A doubled member or nail-laminated beam is required to carry the roof load whether a gable or low-slope roof is used. These beams are made up by using spacers of *lath* or plywood between 2-inch members so that the beam is the same size (3½ inches) as the nominal 4- by 4-inch solid posts used for final support of the roof. End beams made up of doubled 2-inch members are fastened to the outside beam and to the house (fig. 2-104). Thus, the outline of the porch is now formed by the front and end beams. One method of assembling this roof framing is by nailing it together over the floor framing and raising it in place. Temporary 2- by 4-inch or larger braces are used to support this beam framing while the roof is being constructed. Use enough braces to prevent movement. Use joist hangers or a length of angle iron to fasten one end of the beam to the house framing. The beam ends can also be carried through the wall and supported by auxiliary studs when possible. It is often desirable to provide a 6-foot 8-inch or a 7-foot clearance from the porch floor to the bottom of the beams so that standard units can be used if screening or enclosing is planned in the future.

Ceiling joists are now fastened to the beam at one end and to the studs of the house at the other. They should be spaced 16 or 24 inches on center to provide nailing for a ceiling material.

Rafters, either 2- by 4- or 2- by 6-inch in size depending on the span, are now measured and cut at the ridge and wall line as described in the section on pitched roofs. A 1- by 6-inch ridgepole is used to tie each pair of rafters together. Gable end studs are now cut to fit between the end rafters and the outside beam (fig. 2-104). Space them to accommodate the panel siding or other finish. Toenail the studs to the outside beam and to the end rafters.

The roof sheathing, the fly rafters, and the roofing are applied as described in the sections on roof systems and coverings. Use flashing at the junction of the roof and the end wall of the house when applying the roofing.

The matched flooring may now be applied to the floor across the joists. Extend it beyond the outer

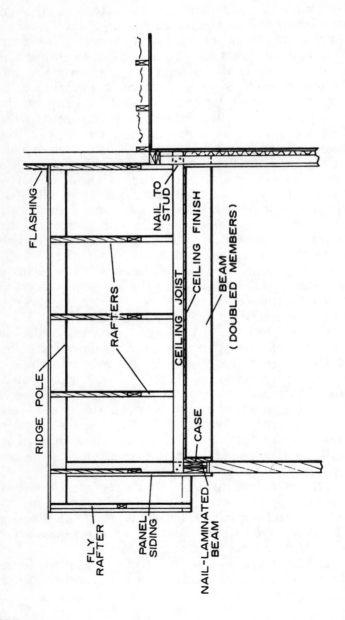

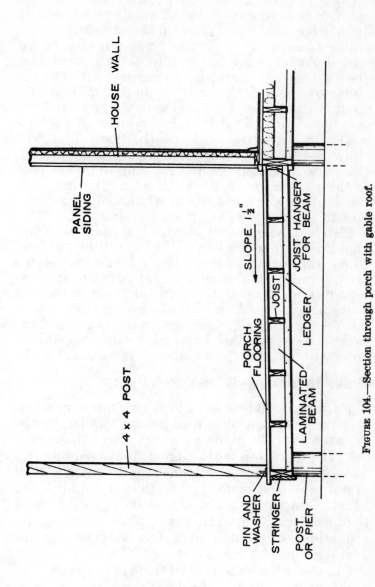

FIGURE 104.—Section through porch with gable roof.

edge (fig. 2-104). Use sevenpenny or eightpenny flooring nails and blind-nail to each joist. It is good practice to apply a saturating coat of water-repellent preservative on the surface and edges. This will provide protection until the floor is painted. Some decay-resistant wood species require little protection other than this treatment.

The nominal 4- by 4-inch posts can now be installed. When they are cut to length, drill and drive a small $3/8$- or $1/2$-inch-diameter pin into the center of one end. A matching hole is drilled into the floor, a mastic calk applied to the area, and the post positioned (fig. 2-104). Use a large galvanized washer between the post and the floor. This will allow moisture to evaporate and prevent decay. Use toenailing and metal strapping to tie the post to the roof framing. In areas of high winds, use a *bolt* or lag screw instead of the pin from the post to some part of the floor framing.

Matched boards, plywood, or similar covering materials can be used to finish the ceiling surface. The exterior and interior of the laminated beams are normally cased with nominal 1-inch boards except where siding at the gable end can be carried over the exposed face of the beam. A 1- by 4-inch member is normally used under the beam and between the posts. All exposed nailing should be done with galvanized or other rust-resistant nails.

Framing for Low-Slope Porch Roof

Framing a flat or low-slope roof for a porch is relatively simple. A nailed beam, similar to the type used for the gable roof, is required to support the ends of rafter-joists (fig. 2-105). Single end members, however, are used to tie the beam to wall framing of the house rather than a doubled end beam. Temporary posts or braces are used to hold the beam framing after it is in place. A minimum roof slope of 1 inch in a foot is desirable for drainage.

Cut the rafters and toenail them to the beam with eightpenny nails and face-nail the opposite ends to the rafter or the ceiling joists of the house (fig. 2-105). When spacing is not the same and members do not join the rafters or ceiling joists of the house, toenail the ends of the porch rafters to the wall plate. Plywood or board sheathing is applied, and when end overhang is desired, a fly rafter is added. When desired, 1- by 6-inch facia boards are applied to the ends of the rafters and to the fly rafter with sevenpenny or eightpenny galvanized siding nails. Roofing is applied as has been recommended for low-slope roofs; underlayment followed by double-coverage surfaced roll roofing. On low slopes, no exposed nails are used. A ribbon of asphalt roof cement or lap seal material is used under the lapped edge. Extend the roofing beyond the facia enough to form a natural drip edge. Ceiling covering and other trim are used in the same manner as for the gable roof.

Steps & Stairs

Outside entry platforms and steps are required for most types of wood-frame houses. Houses constructed over a full basement or crawl space normally require a platform with steps at outside doors. In houses with masonry foundations, these outside entry stoops consist of a concrete perimeter wall with poured concrete steps. However, a well-constructed wood porch will serve as well and normally cost less.

Inside stairs leading to an attic or second-floor bedrooms in a house with a steep roof slope, or to the basement, must be provided for during construction of the house. This includes framing of the floor joists to accommodate the stairway and providing walls and *carriages* for the *treads* and *risers*. Even though the second floor might not be completed immediately, a stairway should be included during construction of such houses.

Outside wood stoops, platforms, and open plank stairs should give satisfactory service if these simple rules are followed: (a) All wood in contact

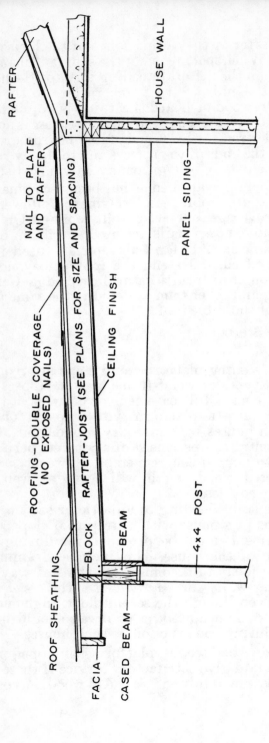

FIGURE 2-105. Section through side porch with low-slope roof.

with or embedded in soil should be pressure-treated, as outlined in the early section on "Post Foundations"; (b) all untreated wood parts should have a 2-inch minimum clearance above the ground; (c) avoid pockets or areas in the construction where water cannot drain away; (d) if possible, use wood having moderate to good decay resistance; (e) use initial and regular applications of water-repellent preservative to exposed untreated wood surfaces; and (f) use vertical grain members.

Wood Stoop—Low Height

A simple all-wood stoop consists of treated posts embedded in the ground, cross or bearing members, and spaced treads. One such design is shown in figure 2-106A. Because this type of stoop is low, it can serve as an entry for exterior doors where the floor level of the house is no more than 24 inches above the ground. Railings are not usually required. The platform should be large enough so that the storm door can swing outward freely. An average size is about 3½ feet deep by 5 feet wide.

Use treated posts of 5- to 7-inch diameter and embed them in the soil at least 3 feet. Nail and bolt (with galvanized fasteners) a crossmember (usually a nominal 2- by 4-inch member) to each side of the posts (fig. 2-106B). The posts snould be faced slightly at these areas. For a small, 3½- by 5-foot stoop, four posts are usually sufficient. Posts are spaced about 4 feet apart across the front.

Supports for the tread of the first step are supplied by pairs of 2- by 4-inch members bolted to the forward posts (fig. 2-106A). The inner ends are blocked with short pieces of 2- by 4-inch members to the upper crosspieces (fig. 2-106A). Treads consist of 2- by 4- or 2- by 6-inch members, spaced about ¼ inch apart. Use two sixteenpenny galvanized plain or ring-shank nails for each piece at each supporting member.

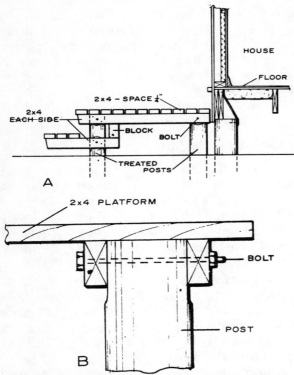

FIGURE 2-106. Low-height wood stoop. A, Side elevation; B, connection to post.

Some species of wood have natural decay resistance and others are benefited by the application of a water-repellent preservative followed by a good deck paint. If desired, a railing can be added by bolting short upright members to the 2- by 4-inch crossmembers. Horizontal railings can be fastened to the uprights.

Wood Stoop—Medium Height

A wood entry platform requiring more than one or two steps is usually designed with a railing and stair stringers. If the platform is about 3½ by 5 feet, two 2- by 12-inch carriages can be used to support the treads (fig. 2-107A). In most cases, the bottoms of the carriages are supported by treated posts embedded in the ground or by an embedded treated timber. The upper ends of the carriages are supported by a 2- by 4-inch ledger fastened to posts and are face-nailed to the platform framing with twelvepenny galvanized nails. The carriage at the house side can be supported in the same way when interior posts are used.

When the platform is narrow, a nominal 3- by 4-inch ledger is fastened to the floor framing of the house with fortypenny galvanized spikes or 5-inch lag screws. The 2- by 6-inch floor planks are nailed to the ledger and to the double 2- by 4-inch beam bolted to the post (fig. 2-107B). Use two sixteenpenny galvanized plain or ring-shank nails for each tread. When a wide platform is desired, an inside set of posts and doubled 2 by 4 or larger beams should be used.

Railings can be made of 2- by 4-inch uprights bolted or lagscrewed to the outside beams. These members are best fastened with galvanized bolts or lag screws. Horizontal railing in 1- by 4- or 1- by 6-inch size can then be fastened to the uprights. When an enclosed skirting is desired, 1- by 4-inch slats can be nailed to the outside of the beam and an added lower nailing member (fig. 2-107A). Treat all exposed untreated wood with a heavy application of water-repellent preservative. When a paint finish is desired, use a good deck paint. See section on "Painting and Finishing" for details.

Inside Stairs

When stairs to a second-floor area are required, the first-floor ceiling joists are framed to accommodate the stairway. When basement stairs are used, the first-floor joists must also be framed for the stairway. Two types of simple stair *runs* are commonly used in a small house, the straight run (fig. 2-108A) and the long L (fig. 2-108B). An open

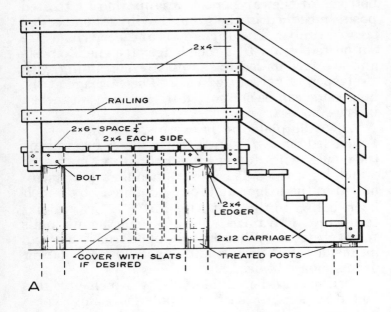
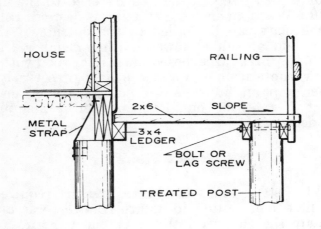

FIGURE 2-107. Medium-height wood stoop. A, Side elevation; B, connection to post.

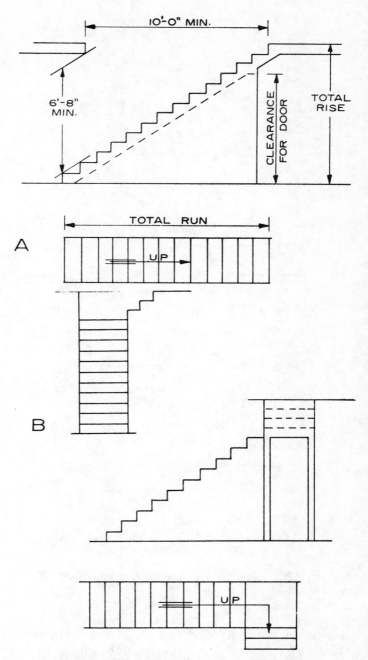

FIGURE 2-108. Stair types. A, Straight run; B, long L.

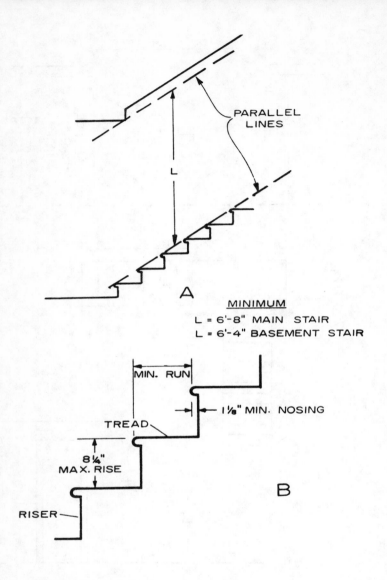

FIGURE 2-109. Stair measurements. A, Head room; B, riser-tread sizes.

length of 10 feet is normally sufficient for adequate headroom with a width of 2½ to 3 feet. A clear width of 2 feet 8 inches is considered minimum for a main stair.

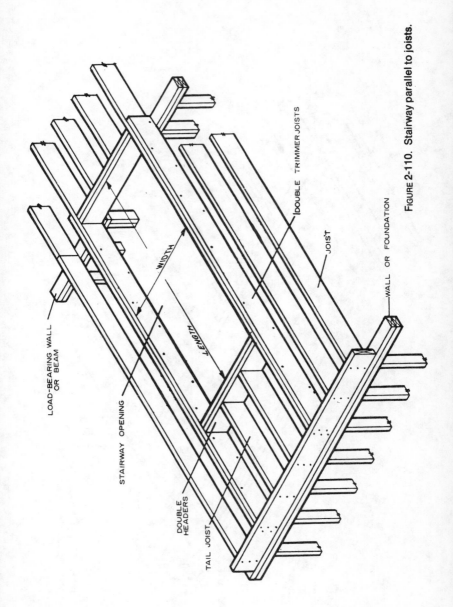

FIGURE 2-110. Stairway parallel to joists.

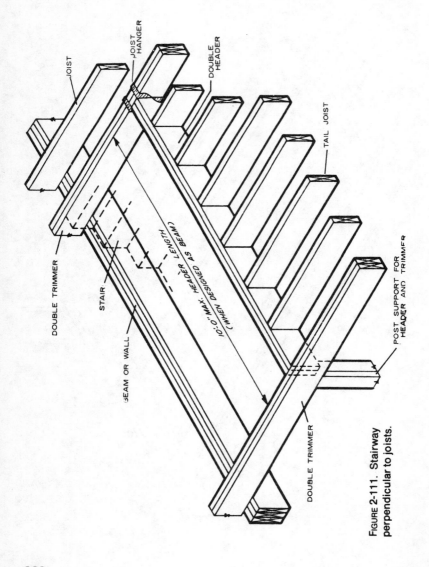

Figure 2-111. Stairway perpendicular to joists.

Two of the most important considerations in the design of inside stairways are headroom and the relation of the riser height to the length of the tread. The minimum headroom for stairs should be 6 feet 4 inches for basements or secondary stairs and 6 feet 8 inches for main stairs fig. 2-109A). The relation of the riser to the run is shown in fig. 2-109B. A good rule of thumb to apply is: The riser times the tread in inches should equal about 75.

When the length of the stairway is parallel to the joists the opening is framed (fig. 2-110). When the stairway is arranged so that the opening is perpendicular to the length of the joists, the framing should follow the details shown in figure 2-111. Nailing and framing should comply with table 1 and the descriptions in the section on "Floor Systems."

The stair carriages are normally made from 2-by 12-inch members. They provide support for the stairs and nailing surfaces for the treads and risers. The carriages can be nailed to a finish stringer and the wall studs behind with a sixteenpenny nail at each stud (fig.2-112A). The carriages can also be mounted directly to the wall studs or over the drywall finish and the finish stringer notched to them (fig.2-112B). When no wall is present to fasten the tops of the carriages in place, use a ledger similar to that shown in figure 107A for an outside stair. The tops of the carriages are notched to fit this ledger. Two carriages are sufficient when treads are at least $1\frac{1}{16}$ inches thick and the stair is less than 2 feet 6 inches wide. Use three carriages when the stair is wider than this. When plank treads $1\frac{5}{8}$ inches thick are used, two carriages are normally sufficient for stair widths up to 3 feet.

After the carriages are mounted to the wall and treads and risers cut to length, nail the bottom riser to each carriage with two eightpenny finish nails. The first tread, if $1\frac{1}{16}$ inches thick, is then nailed to each carriage with two tenpenny finish

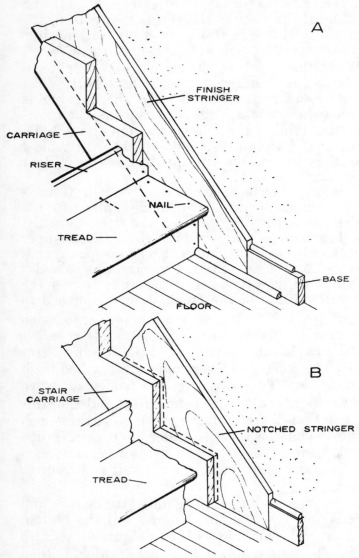

FIGURE 2-112. Stair construction details. A, Full stringer; B, notched stringer.

nails and to the riser below with at least two tenpenny finish nails. Proceed up the stair in this same manner. If 1⅝-inch-thick treads are used, a twelvepenny finish nail may be required. Use three nails at each carriage, but eliminate nailing to the riser below. All finish nails should be set.

Finished stairs with stringers routed to fit the ends of the treads and risers, and with railing and *balusters* or a handrail, are generally used in main stairs to second-floor rooms in moderate-cost houses, but probably should not be considered for low-cost homes.

PAINTING AND FINISHING

The durability of an exterior finish is materially affected by the wood characteristics. Woods that are high in *density* (heavy), such as dense hardwoods, will be more difficult to finish effectively than lightweight woods.

The amount and distribution of summerwood (darker *grained* portions) on the surface of softwood lumber also influence the success of the finishing procedure. Finishes, particularly paints, will last longer on surfaces with a low proportion of summerwood.

The manner in which lumber is sawn from the log influences its finishing characteristics. Paint-type and film-forming finishes always perform best on *vertical-grain* lumber because the summerwood is better distributed on the surface and because vertical-grain lumber is low in swelling.

All woods shrink or swell as they lose or absorb water. Species which shrink and swell the least are best for painting. Checking, warping of wood, and paint peeling are more likely to be critical on woods which are hard, dense, and high in swelling.

Wood that is free of knots, pitch pockets, and other defects is the preferred base for paints, but these defects have little adverse effect on penetrating-type finishes. Smoothly planed surfaces are best for paint finishes, while rougher or sawn surfaces are preferred for penetrating (non-film-forming) finishes.

Exterior Finishes

Unfinished wood.—Permitting the wood to weather naturally without protection of any kind is, of course, very simple and economical. Wood fully exposed to all elements of the weather, rain and sun being the most important, will wear away at the approximate rate of only a quarter of an inch in a century. The time required for wood to weather to the final gray color will depend on the severity of exposure. Wood in protected areas will be much slower to gray than wood fully exposed to the sun on the south side of a building. Early in the graying process, the wood may take on a blotchy appearance because of the growth of micro-organisms on the surface. Migration of wood extractives to the surface also will produce an uneven and unsightly discoloration, particularly in areas that are not washed by rain.

Unfinished lumber will warp more than lumber protected by paint. Warping varies with the density, width, and thickness of the board, basic wood structure, and species. Warp increases with density and the width of the board. To reduce warping to a minimum, best results are obtained when the width of boards does not exceed eight times the thickness. Flat-grain boards warp more than vertical-grain lumber. Baldcypress, the cedars, and redwood are species which have only a slight tendency to warp.

Water-repellent-preservative finishes.—A simple treatment of an exterior wood surface with a water-repellent preservative markedly alters the natural weathering process. Most pronounced is the retention of a uniform natural tan color in the early stages of weathering and a retardation of the uneven graying process which is produced by the growth of mildew on the surface. The water-repellency imparted by the treatment greatly reduces the tendency toward warping, excessive shrinking, and swelling which lead to splitting, and retards the leaching of extractives from the wood and water stain at ends of boards.

This type of finish is quite inexpensive, easily applied, and very easily refinished. Estimated costs for material would be approximately $0.50 to $2 per 100 square feet (spreading rate of 100 to 200 square feet per gallon). Water-repellent preservatives can be applied by brushing, rolling, dipping, and spraying. It is important to thoroughly treat all lap and butt joints and ends of boards. Many brands of effective water-repellent preservatives are on the market. A saving could be made by making a solution from the following components:

Penta concentrate (10:1)	2 quarts.
Boiled linseed oil	1.75 quarts.
Paraffin wax	0.25 to 0.50 pound.
Mineral spirits, turpentine, No. 1 or No. 2 fuel oil.	4 gallons.

Color pigments can also be added to this type of finish. Mix 2 to 6 fluid ounces of colors to each gallon of water-repellent preservative.

The initial applications may be short-lived (1 year), especially in humid climates and on species that are susceptible to mildew, such as sapwood and certain hardwoods. Under more favorable conditions, such as on rough cedar surfaces which will absorb large quantities of the solution, the finish will last more than 2 years.

When blotchy discolorations of mildew start to appear on the wood, re-treat the surface with water-repellent-preservative solution. If extractives have accumulated on the surface in protected areas, clean these areas by mild scrubbing with a detergent or trisodium-phosphate solution.

The continued use of these water-repellent-preservative solutions will effectively prevent serious decay in wood in above-ground installation. This finishing method is recommended for all wood species and surfaces exposed to the weather.

Only aluminum or stainless steel nails will prevent discoloration on the siding. Galvanized nails will show light stains after several years. Steel nails without rust-resistance treatment should not be used.

Penetrating pigmented stain finishes.—The penetrating stains also are effective and economical finishes for all kinds of lumber and plywood surfaces, especially those that are rough-sawn, weathered, and textured. Knotty wood boards and other lower quality grades of wood which would be difficult to paint also can be finished successfully with penetrating stains.

These stains penetrate into the wood without forming a continuous film on the surface. Because there is no film or coating, there can be no failure by cracking, peeling, and blistering. Stain finishes are easily prepared for refinishing and easily maintained.

The penetrating pigmented stains form a flat and semitransparent finish. They permit only part of the wood-grain pattern to show through. A variety of colors can be achieved with finish. Shades of brown, green, red, and gray are possible. The only color which is not available is white. This color can be provided only through the use of white paint.

Stains are quite inexpensive and easy to apply. To avoid the formation of lap marks, the entire length of a course of siding should be finished without stopping. Only one coat is recommended on smoothly planed surfaces, where it will last 2 to 3 years. After refinishing, however, the second coat will last 6 to 7 years because the weathered surface has adsorbed more of the stain than the smoothly planed surface.

Two-coat staining is possible on rough-sawn or weathered surfaces, but both coats should be applied within a few hours of each other. When using a two-coat system, the first coat should never be allowed to dry before the second is applied, because this will seal the surface and prevent the second coat from penetrating. A finish life of up to 10 years can be achieved when two coats are applied to a rough or weathered surface.

Very satisfactory penetrating stains can be prepared by home-mixing the following ingredients:

Boiled linseed oil	3 gallons.
Penta concentrate	½ gallon.
Paraffin wax	½ pound.
Colors-in-oil (tinting colors)	1 quart.
Paint thinner	1 gallon.

Boiled linseed oil is available in most paint stores and mail-order houses. Penta concentrate is a common name for a solution of pentachlorophenol which is about a 40 percent concentration and is also known as a 10 to 1 concentrate. It is available from several manufacturers by mail order. Paraffin wax, used to seal jelly glasses, can be bought from local grocery stores. Colors-in-oil or tinting colors are available from paint stores and artists' supply stores. Paint thinners which can be used are mineral spirits or *turpentine*. In areas where available, No. 1 or No. 2 fuel oil also can be used. In warm moist areas which enhance fungal growth, the penta content should be doubled.

All ingredients will go into solution quite easily if temperatures are 70° F. or above. Dissolving the wax, which is the most difficult step in the process, can be aided by cutting it into fine chips or by melting in a double boiler before adding to other ingredients. Allow the solution to stand overnight and occasionally stir vigorously during use to keep pigments uniformly suspended.

CAUTION: Turpentine, mineral spirits, and other paint thinners are volatile flammable solvents. Their concentrated vapors should not be breathed or exposed to sparks or flames that can ignite them. It is safer to mix ingredients outdoors or in an open garage than in a closed room in a house.

Stained surface should be refinished only when the colors fade and bare wood is beginning to show. A light steel-wooling and hosing with water to remove surface dirt and mildew are all that are needed to prepare the surface. Restain after the surfaces have thoroughly dried.

Clear film finishes.—Clear finishes based on *varnish*, which form a coating or film on the surface, should not be used on wood exposed fully to the weather. These finishes are quite expensive and often begin to deteriorate within 1 year. Refinishing is a frequent, difficult, and time-consuming process.

Exterior paints.—Of all the finishes, paints provide the widest selection of color. When properly selected and applied, it will also provide the most protection to wood against weathering. The durability of paint coatings on exterior wood, however, is affected by many variables, and much care is needed in the selection of the wood surface material, type of paint, and method of application to achieve success in painting. The original and maintenance costs are higher for a paint finish than for either the water-repellent-preservative treatment or penetrating-stain finish.

Paint performance is affected by species, density, wood structure, extractives, and defects such as knots and pitch pockets.

Best paint durability will be achieved on the select high grades of vertical-grain western redcedar, redwood, and low-density pines such as the white pines, sugar pine, and ponderosa pine. Exterior-grade plywood which has been overlaid with medium-density resin-treated paper is another wood-base material on which paint will perform very well.

Follow these three simple steps when painting wood:

> (1) Apply water-repellent preservative to all joints by brushing or spraying. Treat all lap and butt joints, ends, and edges of lumber, and window sash and trim. Allow 2 warm days of drying before painting.

(2) Prime the treated wood surface with an oil-base paint free of zinc-oxide pigment. Do *not* use a porous blister-resistant paint as primer on wood surfaces. Apply sufficient primer so the grain of the wood cannot be seen. Open joints should be calked after priming.

(3) Apply two topcoats of high-quality oil, alkyd, or latex paint over the primer. Two topcoats, especially, should be used on the south side, which has the most severe exposure.

Interior Finishes

Interior finishes for wood and dry-wall or plaster surfaces are usually intended to serve one or more of the following purposes:

(1) Make the surface easy to clean.
(2) Enhance the natural beauty of wood.
(3) Achieve a desired color decor.
(4) Impart wear resistance.

The type of finish depends largely upon type of area and the use to which the area will be put. The various interior areas and finish systems employed in each are summarized in table 3. Wood surfaces can be finished either with a clear finish or a paint. Plaster-base materials are painted.

Wood Floors

Hardwood floors of oak, birch, beech, and maple are usually finished by applying two coats of wood seal, also called floor seal, with light sanding between coats. A final coat of paste wax is then applied and buffed. This finish is easily maintained by rewaxing. The final coat can also be a varnish instead of a sealer. The varnish finishes are used when a high gloss is desired.

When floors are to be painted, an *undercoater* is used, and then at least one topcoat of floor and deck enamel is applied.

TABLE 3.—*What interior finish to use—and where*

	Primer or under- coater	Rubber latex	Flat oil paint	Semi- gloss paint	Floor (wood) seal [1]	Var- nish [1]	Floor or deck enamel
Wood floors	X				X	X	X
Wood paneling and trim	X	X	X	X	X	X	
Kitchen and bathroom walls	X			X			
Dry wall and plaster	X	X	X				

[1] Paste wax can be applied over floor seal and varnish base.

Glossary of Housing Terms

Airway. A space between roof insulation and roof boards for movement of air.

Apron. The flat member of the inside trim of a window placed against the wall immediately beneath the stool.

Asphalt. Most native asphalt is a residue from evaporated petroleum. It is insoluble in water but soluble in gasoline and melts when heated. Used widely in building for such items as waterproof roof coverings of many types, exterior wall coverings, and flooring tile.

Attic ventilators. In houses, screened openings provided to ventilate an attic space. They are located in the soffit area as inlet ventilators and in the gable end or along the ridge as outlet ventilators. They can also consist of powerdriven fans used as an exhaust system. See also *Louver*.

Backfill. The replacement of excavated earth into a trench or pier excavation around and against a basement foundation.

Balusters. Usually small vertical members in a railing used between a top rail and the stair treads or a bottom rail.

Base or baseboard. A board placed around a room against the wall next to the floor to finish properly between floor and plaster or dry wall.

Base molding. Molding used to trim the upper edge of interior baseboard.

Base shoe. Molding used next to the floor on interior baseboard. Sometimes called a carpet strip.

Batten. Narrow strips of wood used to cover joints or as decorative vertical members over plywood or wide boards.

Beam. A structural member transversely supporting a load.

Bearing partition. A partition that supports any vertical load in addition to its own weight.

Bearing wall. A wall that supports any vertical load in addition to its own weight.

Bed molding. A molding in an angle, as between the overhanging cornice, or eaves, of a building and the sidewalls.

Blind-nailing. Nailing in such a way that the nailheads are not visible on the face of the work. Usually at the tongue of matched boards.

Blind stop. A rectangular molding, usually 3/4 by 1 3/8 inches or more in width, used in the assembly of a window frame. Serves as a stop for storm and screen or combination windows and to resist air infiltration.

Boiled linseed oil. Linseed oil in which enough lead, manganese, or cobalt salts have been incorporated to make the oil harden more rapidly when spread in thin coatings.

Bolts, anchor. Bolts to secure a wooden sill plate to concrete or masonry floor or wall or pier.

Boston ridge. A method of applying asphalt or wood shingles at the ridge or at the hips of a roof as a finish.

Brace. An inclined piece of framing lumber applied to wall or floor to stiffen the structure. Often used on walls as temporary bracing until framing has been completed.

Buck. Often used in reference to rough frame opening members. Door bucks used in reference to metal door frame.

Built-up roof. A roofing composed of three to five layers of asphalt felt laminated with coal tar, pitch, or asphalt. The top is finished with crushed slag or gravel. Generally used on flat or low-pitched roofs.

Butt joint. The junction where the ends of two timbers or other members meet in a square-cut joint.

Cabinet. A shop- or job-built unit for kitchens or other rooms. Often includes combinations of drawers, doors, and the like.

Casing. Molding of various widths and thicknesses used to trim door and window openings at the jambs.

Casement frames and sash. Frames of wood or metal enclosing part or all of the sash, which may be opened by means of hinges affixed to the vertical edges.

Collar beam. Nominal 1- or 2-inch-thick members connecting opposite roof rafters. They serve to stiffen the roof structure.

Combination doors or windows. Combination doors or windows used over regular openings. They provide winter insulation and summer protection. They often have self-storing or removable glass and screen inserts. This eliminates the need for handling a different unit each season.

Concrete, plain. Concrete without reinforcement, or reinforced only for shrinkage or termperature changes.

Condensation. Beads or drops of water, and frequently frost in extremely cold weather, that accumulate on the inside of the exterior covering of a building when warm, moisture-laden air from the interior reaches a point where the temperature no longer permits the air to sustain the moisture it holds. Use of louvers or attic ventilators will reduce moisture condensation in attics. A vapor barrier under the gypsum lath or dry wall on exposed walls will reduce condensation in walls.

Conduit, electrical. A pipe, usually metal, in which wire is installed.

Construction, dry-wall. A type of construction in which the interior wall finish is applied in a dry condition, generally in the form of sheet materials or wood paneling, as contrasted to plaster.

Construction, frame. A type of construction in which the structural parts are of wood or depend upon a wood frame for support. In building codes, if masonry veneer is applied to the exterior walls, the classification of this type of construction is usually unchanged.

Coped joint. Fitting woodwork to an irregular surface. In moldings, cutting the end of one piece to fit the molded face of the other at an interior angle to replace a miter joint.

Corner bead. A strip of formed sheet metal, sometimes combined with a strip of metal lath, placed on corners before plastering to reinforce them. Also, a strip of wood finish three-quarters round or angular placed over a plastered corner for protection.

Corner boards. Used as trim for the external corners of a house or other frame structure against which the ends of the siding are finished.

Corner braces. Diagonal braces at the corners of frame structure to stiffen and strengthen the wall.

Cornice. Overhang of a pitched roof at the eave line, usually consisting of a facia board, a soffit for a closed cornice, and appropriate moldings.

Counterflashing. A flashing usually used on chimneys at the roofline to cover shingle flashing and to prevent moisture entry.

Cove molding. A molding with a concave face used as trim or to finish interior corners.

Crawl space. A shallow space below the living quarters of a basementless house, sometimes enclosed.

d. See *Penny.*

Dado. A rectangular groove across the width of a board or plank. In interior decoration, a special type of wall treatment.

Deck paint. An enamel with a high degree of resistance to mechanical wear, designed for use on such surfaces as porch floors.

Density. The mass of substance in a unit volume. When expressed in the metric system (in g. per cc.), it is numerically equal to the specific gravity of the same substance.

Dimension. See *Lumber, dimension.*

Doorjamb, interior. The surrounding case into and out of which a door closes and opens. It consists of two upright pieces, called side jambs, and a horizontal head jamb.

Dormer. A projection in a sloping roof, the framing of which forms a vertical wall suitable for windows or other openings.

Downspout. A pipe, usually metal, for carrying rainwater from roof gutters.

Dressed and matched (*tongued and grooved*). Boards or planks machined in such a manner that there is a groove on one edge and a corresponding tongue on the other.

Drier, paint. Usually oil-soluble soaps of such metals as lead, manganese, or cobalt, which, in small proportions, hasten the oxidation and hardening (drying) of the drying oils in paints.

Drip cap. A molding placed on the exterior top side of a door or window frame to cause water to drip beyond the outside of the frame.

Dry-wall. see *Construction, dry wall.*

Ducts. In a house, usually round or rectangular metal pipes for distributing warm air from the heating plant to rooms, or air from a conditioning device, or as cold air returns. Ducts are also made of asbestos and composition materials.

Eaves. The overhang of a roof projecting over the walls.

Face nailing. To nail perpendicular to the initial surface or to the junction of the pieces joined.

Facia or fascia. A flat board, band, or face, used sometimes by itself but usually in combination with moldings, often located at the outer face of the cornice.

Flashing. Sheet metal or other material used in roof and wall construction to protect a building from seepage of water.

Flat paint. An interior paint that contains a high proportion of pigment, and dries to a flat or lusterless finish.

Flue. The space or passage in a chimney through which smoke, gas, or fumes ascend. Each passage is called a flue, which, together with any others and the surrounding masonry, make up the chimney.

Flue lining. Fire clay or terra-cotta pipe, round or square, usually made in all of the ordinary flue sizes and in 2-foot lengths, used for the inner lining of chimneys with a brick or masonry work around the outside. Flue lining in chimneys runs from about a foot below the flue connection to the top of the chimney.

Fly rafter. End rafters of the gable overhang supported by roof sheathing and lookouts.

Footing. A masonry section, usually concrete in a rectangular form wider than the bottom of the foundation wall or pier it supports.

Foundation. The supporting portion of a structure below the first-floor construction, or below grade, including the footings.

Framing, balloon. A system of framing a building in which all vertical structural elements of the bearing walls and partitions consist of single pieces extending from the top of the foundation sill plate to the roofplate and to which all floor joists are fastened.

Framing, platform. A system of framing a building in which floor joists of each story rest on the top plates of the story below or on the foundation sill for the first story, and the bearing walls and partitions rest on the subfloor of each story.

Frieze. In house construction, a horizontal member connecting the top of the siding with the soffit of the cornice or roof sheathing.

Frostline. The depth of frost penetration in soil. This depth varies in different parts of the country. Footings should be placed below this depth to prevent movement.

Furring. Strips of wood or metal applied to a wall or other surface to even it and usually to serve as a fastening base for finish material.

Gable. The triangular vertical end of a building formed by the eaves and ridge of a sloped roof.

Gloss (paint or enamel). A paint or enamel that contains a relatively low proportion of pigment and dries to a sheen or luster.

Girder. A large or principal beam of wood or steel used to support concentrated loads at isolated points along its length.

Grain. The direction, size, arrangement, appearance, or quality of the fibers in wood.

Grain, edge (vertical). Edge-grain lumber has been sawed parallel to the pith of the log and approximately at right angles to the growth rings; i.e., the rings form an angle of 45° or more with the surface of the piece.

Gusset. A flat wood, plywood, or similar type member used to provide a connection at the intersection of wood members. Most commonly used at joints of wood trusses. They are fastened by nails, screws, bolts, or adhesives.

Gutter or eave trough. A shallow channel or conduit of metal or wood set below and along the eaves of a house to catch and carry off rainwater from the roof.

Header. (a) A beam placed perpendicular to joists and to which joists are nailed in framing for chimney, stairway, or other opening. (b) A wood lintel.

Heartwood. The wood extending from the pith to the sapwood, the cells of which no longer participate in the life processes of the tree.

Hip. The external angle formed by the meeting of two sloping sides of a roof.

Hip roof. A roof that rises by inclined planes from all four sides of a building.

Insulation board, rigid. A structural building board made of wood or cane fiber in ½- and $^{25}/_{32}$-inch thicknesses. It can be obtained in various size sheets, in various densities, and with several treatments.

Insulation, thermal. Any material high in resistance to heat transmission that, when placed in the walls, ceilings, or floors of a structure, will reduce the rate of heat flow.

Jack rafter. A rafter that spans the distance from the wallplate to a hip, or from a valley to a ridge.

Jamb. The side and head lining of a doorway, window, or other opening.

Joint. The space between the adjacent surfaces of two members or components joined and held together by nails, glue, cement, mortar, or other means.

Joint cement. A powder that is usually mixed with water and used for joint treatment in gypsum-wallboard finish. Often called "spackle."

Joist. One of a series of parallel beams, usually 2 inches thick, used to support floor and ceiling loads, and supported in turn by larger beams, girders, or bearing walls.

Knot. In lumber, the portion of a branch or limb of a tree that appears on the edge or face of the piece.

Landing. A platform between flights of stairs or at the termination of a flight of stairs.

Lath. A building material of wood, metal, gypsum, or insulating board that is fastened to the frame of a building to act as a plaster base.

Ledger strip. A strip of lumber nailed along the bottom of the side of a girder on which joists rest.

Light. Space in a window sash for a single pane of glass. Also, a pane of glass.

Lintel. A horizontal structural member that supports the load over an opening such as a door or window.

Lookout. A short wood bracket or cantilever to support an overhanging portion of a roof or the like, usually concealed from view.

Louver. An opening with a series of horizontal slats so arranged as to permit ventilation but to exclude rain, sunlight, or vision. See also *Attic ventilators.*

Lumber. Lumber is the product of the sawmill and planing mill not further manufactured other than by sawing, resawing, and passing lengthwise through a standard planing machine, cross cutting to length, and matching.

Lumber, boards. Yard lumber less than 2 inches thick and 2 or more inches wide.

Lumber, dimension. Yard lumber from 2 inches to, but not including, 5 inches thick, and 2 or more inches wide. Includes joists, rafters, studs, plank, and small timbers. The actual size dimension of such lumber after shrinking from green dimension and after machining to size or pattern is called the dress size.

Lumber, matched. Lumber that is dressed and shaped on one edge in a grooved pattern and on the other in a tongued pattern.

Lumber, shiplap. Lumber that is edge-dressed to make a close rabbeted or lapped joint.

Lumber, yard. Lumber of those grades, sizes, and patterns which are generally intended for ordinary construction, such as framework and rough coverage of houses.

Masonry. Stone, brick, concrete, hollow-tile, concrete-block, gypsum-block, or other similar building units or materials or a combination of the same, bonded together with mortar to form a wall, pier, buttress, or similar mass.

Meeting rails. Rails sufficiently thicker than a window to fill the opening between the top and bottom sash made by the parting stop in the frame of double-hung windows. They are usually beveled.

Millwork. Generally all building materials made of finished wood and manufactured in millwork plants and planing mills are included under the term "millwork." It includes such items as inside and outside doors, window and doorframes, blinds, porchwork, mantels, panelwork, stairways, moldings, and interior trim. It normally does not include flooring, ceiling, or siding.

Miter joint. The joint of two pieces at an angle that bisects the joining angle. For example, the miter joint at the side and head casing at a door opening is made at a 45° angle.

Moisture content of wood. Weight of the water contained in the wood, usually expressed as a percentage of the weight of the ovendry wood.

Mortise. A slot cut into a board, plank, or timber, usually edgewise, to receive tenon of another board, plank, or timber to form a joint.

Molding. A wood strip having a curved or projecting surface used for decorative purposes.

Natural finish. A transparent finish which does not seriously alter the original color or grain of the natural wood. Natural finishes are usually provided by sealers, oils, varnishes, water-repellent preservatives, and other similar materials.

Nonloadbearing wall. A wall supporting no load other than its own weight.

Notch. A crosswise rabbet at the end of a board.

O.C., on center. The measurement of spacing for studs, rafters, joists, and the like in a building from center of one member to the center of the next.

Paint. A combination of pigments with suitable thinners or oils to provide decorative and protective coatings.

Panel. In house construction, a thin flat piece of wood, plywood, or similar material, framed by stiles and rails as in a door or fitted into grooves of thicker material with molded edges for decorative wall treatment.

Paper, sheathing or building. A building material, generally paper or felt used in wall and roof construction as a protection against the passage of air and sometimes moisture.

Parting stop or strip. A small wood piece used in the side and head jambs of double-hung windows to separate upper and lower sash.

Partition. A wall that subdivides spaces within any story of a building.

Penny. As applied to nails, it originally indicated the price per hundred. The term now serves as a measure of nail length and is abbreviated by the letter *d*.

Perm. A measure of water vapor movement through a material (grains per square foot per hour per inch of mercury difference in vapor pressure).

Pier. A column of masonry, usually rectangular in horizontal cross section, used to support other structural members.

Pigment. A powdered solid in suitable degree of subdivision for use in paint or enamel.

Pitch. The incline slope of a roof, or the ratio of the total rise to the total width of a house; i.e., an 8-foot rise and a 24-foot width are a ⅓ pitch roof. *Roof slope* is expressed in inches of rise per 12 inches of run.

Plate. Sill plate: a horizontal member anchored to a masonry wall. Sole plate: bottom horizontal member of a frame wall. Top plate: top horizontal member of a frame wall supporting ceiling joists, rafters, or other members.

Plumb. Exactly perpendicular; vertical.

Plywood. A piece of wood made of three or more layers of veneer joined with glue and usually laid with the grain of adjoining plies at right angles. Almost always an odd number of plies are used to provide balanced construction.

Porch. A roofed area extending beyond the main house. May be open or enclosed and with concrete or wood frame floor system.

Preservative. Any substance that, for a reasonable length of time, will prevent the action of wood-destroying fungi, borers of various kinds, and similar destructive life when the wood has been properly coated or impregnated with it.

Primer. The first coat of paint in a paint job that consists of two or more coats; also the paint used for such a first coat.

Putty. A type of cement usually made of whiting and boiled linseed oil, beaten or kneaded to the consistency of dough, and used in sealing glass in sash, filling small holes and crevices in wood, and for similar purposes.

Quarter round. A small molding that has the cross section of a quarter circle.

Rafter. One of a series of structural members of a roof designed to support roof loads. The rafters of a flat roof are sometimes called roof joists.

Rafter, hip. A rafter that forms the intersection of an external roof angle.

Rafter, valley. A rafter that forms the intersection of an internal roof angle. The valley rafter is normally made of doubled 2-inch-thick members.

Rail. Cross members of panel doors or of a sash. Also the upper and lower members of a balustrade or staircase extending from one vertical support, such as a post, to another.

Rake. The inclined edge of a gable roof (the trim member is a rake molding).

Ridge. The horizontal line at the junction of the top edges of two sloping roof surfaces.

Ridge board. The board placed on edge at the ridge of the roof into which the upper ends of the rafters are fastened.

Rise. In stairs, the vertical height of a step or flight of stairs.

Riser. Each of the vertical boards closing the spaces between the treads of stairways.

Roll roofing. Roofing material, composed of fiber and saturated with asphalt, that is supplied in rolls containing 108 square feet in 36-inch widths. It is generally furnished in weights of 45 to 90 pounds per roll.

Roof sheathing. The boards or sheet material fastened to the roof rafters on which the shingle or other roof covering is laid.

Routed. See *Mortised.*

Run. In stairs, the net width of a step or the horizontal distance covered by a flight of stairs.

Sash. A single light frame containing one or more lights of glass.

Saturated felt. A felt which is impregnated with tar or asphalt.

Scab. A short piece of wood or plywood fastened to two abutting timbers to splice them together.

Sealer. A finishing material, either clear or pigmented, that is usually applied directly over uncoated wood for the purpose of sealing the surface.

Semigloss paint or enamel. A paint or enamel made with a slight insufficiency of nonvolatile vehicle so that its coating, when dry, has some luster but is not very glossy.

Shake. A thick handsplit shingle, resawed to form two shakes; usually edge grained.

Sheathing. The structural covering, usually wood boards or plywood, used over studs or rafters of a structure. Structural building board is normally used only as wall sheathing.

Sheathing paper. See *Paper, sheathing.*

Shingles. Roof covering of asphalt, asbestos, wood, tile, slate, or other material cut to stock lengths, widths, and thicknesses.

Shingles, siding. Various kinds of shingles, such as wood shingles or shakes and nonwood shingles, that are used over sheathing for exterior sidewall covering of a structure.

Shiplap. See *Lumber, shiplap.*

Siding. The finish covering of the outside wall of a frame building, whether made of horizontal weatherboards, vertical boards with battens, shingles, or other material.

Siding, bevel (lap siding). Wedge-shaped boards used as horizontal siding in a lapped pattern. This siding varies in butt thickness from ½ to ¾ inch and in widths up to 12 inches. Normally used over some type of sheathing.

Siding, drop. Usually ¾ inch thick and 6 and 8 inches in width with tongued-and-grooved or shiplap edges. Often used as siding without sheathing in secondary buildings.

Siding, panel. Large sheets of plywood or hardboard which serve as both sheathing and siding.

Sill. The lowest member of the frame of a structure, resting on the foundation and supporting the floor joists or the uprights of the wall. The member forming the lower side of an opening, as a door sill, window sill, etc.

Soffit. Usually the underside covering of an overhanging cornice.

Soil cover (ground cover). A light covering or plastic film, roll roofing, or similar material used over the soil in crawl spaces of buildings to minimize moisture permeation of the area.

Soil stack. A general term for the vertical main of a system of soil, waste, or vent piping.

Sole or sole plate. See *Plate*.

Span. The distance between structural supports such as walls, columns, piers, beams, girders, and trusses.

Square. A unit of measure—100 square feet—usually applied to roofing material. Sidewall coverings are sometimes packed to cover 100 square feet and are sold on that basis.

Stain, shingle. A form of oil paint, very thin in consistency, intended for coloring wood with rough surfaces, like shingles, without forming a coating of significant thickness or gloss.

Stair carriage. Supporting member for stair treads. Usually a 2-inch plank notched to receive the treads; sometimes termed a "rough horse."

Stool. A flat molding fitted over the window sill between jambs and contacting the bottom rail of the lower sash.

Storm sash or storm window. An extra window usually placed on the outside of an existing window as additional protection against cold weather.

Story. That part of a building between any floor and the floor or roof next above.

String, stringer. A timber or other support for cross members in floors or ceilings. In stairs, the support on which the stair treads rest; also stringboard.

Stud. One of a series of slender wood or metal vertical structural members placed as supporting elements in walls and partitions. (Plural: studs or studding.)

Subfloor. Boards or plywood laid on joists over which a finish floor is to be laid.

Tail beam. A relatively short beam or joist supported in a wall on one end and by a header at the other.

Termites. Insects that superficially resemble ants in size, general appearance, and habit of living in colonies; hence, frequently called "white ants." Subterranean termites *do not* establish themselves in buildings by being carried in with lumber, but by entering from ground nests after the building has been constructed. If unmolested, they eat out the woodwork, leaving a shell of sound wood to conceal their activities, and damage may proceed so far so to cause collapse of parts of a structure before discovery. There are about 56 species of termites known in the United States; but the two main species, classified from the manner in which they attack wood, subterranean (ground-inhabiting) termites, the most common, and dry-wood termites, found almost exclusively along the extreme southern border and the Gulf of Mexico in the United States.

Termite shield. A shield, usually of noncorrodible metal, placed in or on a foundation wall or other mass of masonry or around pipes to prevent passage of termites.

Threshold. A strip of wood or metal with beveled edges used over the finished floor and the sill of exterior doors.

Toenailing. To drive a nail at a slant with the initial surface in order to permit it to penetrate into a second member.

Tread. The horizontal board in a stairway on which the foot is placed.

Trim. The finish materials in a building, such as moldings, applied around openings (window trims, door trim) or at the floor and ceiling of rooms (baseboard, cornice, picture molding).

Trimmer. A beam or joist to which a header is nailed in framing for a chimney, stairway, or other opening.

Truss. A frame or jointed structure designed to act as a beam of long span, while each member is usually subjected to longitudinal stress only, either tension or compression.

Turpentine. A volatile oil used as a thinner in paints and as a solvent in varnishes. Chemically, it is a mixture of terpenes.

Undercoat. A coating applied prior to the finishing or top coats of a paint job. It may be the first of two or the second of three coats. In some usage of the word, it may become synonymous with priming coat.

Vapor barrier. Material used to retard the movement of water vapor into walls and prevent condensation in them. Usually considered as having a perm value of less than 1.0. Applied separately over the warm side of exposed walls or as a part of batt or blanket insulation.

Varnish. A thickened preparation of drying oil or drying oil and resin suitable for spreading on surfaces to form continuous, transparent coatings, or for mixing with pigments to make enamels.

Water-repellent preservative. A liquid designed to penetrate into wood and impart water repellency and a moderate preservative protection. It is used for millwork, such as sash and frames, and is usually applied by dipping.

Weatherstrip. Narrow or jamb-width sections of thin metal or other material to prevent infiltration of air and moisture around windows and doors.

Wood Paneling and Trim

Wood trim and paneling are most commonly finished with a clear wood sealer or a stain-sealer combination and then topcoated after sanding with at least one additional coat of sealer or varnish. The final coat of sealer or varnish can also be covered with a heavy coat of paste wax to produce a surface which is easily maintained by rewaxing. Good depth in a clear finish can be achieved by finishing first with one coat of a high-gloss varnish followed with a final coat of semigloss varnish.

Wood trim of nonporous species such as pine can also be painted by first applying a coat of primer or undercoater, followed with a coat of latex, flat, or *semigloss oil-base paint.* Semigloss and *gloss paints* are more resistant to soiling and more easily cleaned by washing than the flat oil and latex paints. Trim of porous wood species such as oak and mahogany requires filling before painting.

Kitchen and Bathroom Walls

Kitchen and bathroom walls, which normally are plaster or dry-wall construction, are finished best with a coat of undercoater and two coats of *semigloss enamel.* This type of finish wears well, is easy to clean, and is quite resistant to moisture.

Dry Wall and Plaster

Plaster and dry-wall surfaces, which account for the major portion of the interior area, are finished with two coats of either *flat oil* or latex paint. An initial treatment with size or sealer will improve holdout (reduce penetration of succeeding coats) and thus reduce the quantity of paint required for good coverage.

FOUNDATION ENCLOSURES

A treated wood post foundation is normally constructed with a crawl space having 18- to 24-inch clearance. This access space may be used for placing floor insulation, for installing a vapor barrier soil cover, for examining and treating soil in termite areas, and for other needs. In colder climates, it is often desirable to enclose a crawl space by some low-cost means, even though the floor is insulated. This is commonly done by fastening skirt boards of a long-lasting sheet material to the outside beams or floor framing. Enclosed crawl spaces should have a soil cover and a small amount of ventilation to assure satisfactory performance.

When a treated post foundation is not used, a masonry wall of concrete or concrete block construction will provide a satisfactory enclosure. Unlike the skirtboard enclosure, the masonry wall normally acts as a support for the floor framing system. Ventilation and the use of a soil cover are also required for a masonry foundation. The soil cover can consist of a 4-mil polyethylene or similar material placed over the soil under the house. Its use reduces ground moisture movement to wood members, which could result in excessive moisture and condensation. Ventilation can consist of standard foundation vents made for this purpose.

Masonry House Foundations

Although a treated wood post foundation will reduce the cost of a crawl-space house substantially, local conditions or a personal choice may indicate the use of a full masonry wall. This type of foundation, like the treated post type, will often require only a minimum amount of grading. Concrete block or poured concrete walls and piers with appropriate footings are accepted methods of providing supporting walls for the floor framing. Their use normally eliminates the need for outer beams as the joists are supported by the walls. Only a center flush or drop beam is required. However, the concrete block wall introduces the need for masonry work, which in some areas may be difficult to obtain.

The perimeter layout for the masonry foundation can be made in the same manner as has been outlined for the wood post foundation system (fig. 2-4). The outside line of the masonry walls will be the same as the outside line of the floor framing. The centerline of the interior load-bearing masonry piers is normally the same as for the posts of the wood foundation. Figure 2-113 is the plan of a typical masonry foundation for a crawl-space house. Footings are required for the outside walls as well as for the masonry piers which support the center beam. A soil cover of polyethylene or similar vapor barrier material should be used in all enclosed crawl-space houses. This prevents ground moisture from moving into the crawl-space area. Uncovered ground can result in high moisture content of floor framing, insulation, and other materials in the crawl space. A poured concrete wall requires some type of formwork while concrete block wall and piers are laid up directly on the footings. The information on concrete and the proper mortar mix for the concrete block is outlined in the chapter on "Foundation Systems" in this manual.

The spacing of the center piers will depend on the size of the center beams. Longer distances between piers will require larger or deeper beams than moderate spans of 8 to 10 feet. These details are shown on the working drawings for each house.

Footings

Footing size is determined by the thickness of the foundation wall. A rule of thumb which is

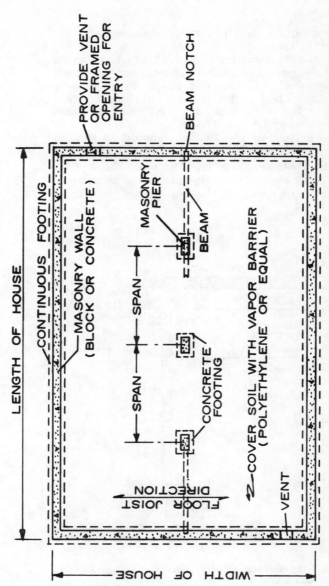

Figure 2-113. Typical masonry foundation wall (concrete block or poured concrete for crawl-space house).

often used for small wood-frame houses under normal soil conditions is: The footing depth should equal the wall thickness and the footing width should be twice the wall thickness. Thus, an 8-inch masonry wall will require a 16- by 8-inch footing (fig. 2-114). The foundation plan will show the footing size for each house. Unusual soil conditions will often require special footing design.

The bottom of the footings should be located below frost line. This may be 4 feet or more in the Northern States. Local regulations or the footing details of neighboring well-constructed houses will indicate this depth. If the soil is stable, no forms

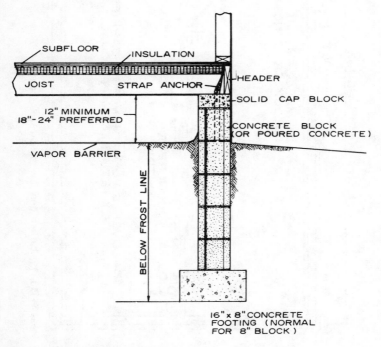

FIGURE 2-114. Section through exterior masonry wall (for crawl-space house).

are required for the sides of the footing trench.

One of the important factors in footing construction is to have the top level all around, especially when concrete block construction is to be used. Drive elevation stakes around the perimeter of the footing so that they can be used as guides when pouring the concrete. These elevations can be established by measuring down from the leveling line described in the chapter on "Foundation Systems." Concrete for footings should be poured over undisturbed soil.

A concrete block wall should normally be finished with a 4-inch solid cap block at the top to provide a good bearing surface for the joists, headers, and stringers(fig. 2-114). Anchor straps for the perimeter joists or headers are desirable in areas where high winds occur. They often consist of perforated or plain 22-gage or heavier galvanized metal straps about 2 inches wide. They are used with a bent "L" shaped base which is placed one or more courses below the cap block. They extend above the top of the wall so that they can be fastened to the edge headers, stringers, or joists. Space them about 8 feet apart and fasten by nailing (fig. 2- 114).

Center Beams

The height of the center masonry piers with relation to the wall height is determined by the type of center beam used. The flush beam uses ledgers to support the joists, but when a drop beam is used the joists bear directly on the top surface. Each system requires a different depth notch for the masonry end walls. This is to assure that the top of all joists, headers, stringers, and the flush beam, when used, have the same elevation.

Flush beam.—The flush beam allows for more clearance in the crawl space as only the amount equal to the depth of the ledgers extends below the bottom of the joists (fig. 2-115A). The joists rest on the ledger and are toenailed to the beam. A strap anchor or a bolt should be used to anchor the beam in high wind areas.

A notch must be provided in the end walls for beam support (fig. 2-115B). The depth of this notch is equal to the depth of the ledgers. Bearing on the wall should be at least 4 inches. A clearance of about ½ inch should be allowed at the sides and ends of the beam. This will provide an airway to prevent the beam from retaining moisture. The size of the notch or beam opening in the wall (when 2- by 4-inch ledgers are used and the beam consists of two nominal 2-inch-thick members) would be approximately:

Length (along length of wall).	7 inches
Width _____	4½ inches (4-inch bearing area, plus clearance)
Height (or depth) _____	3½ inches (or width of ledgers)

The top of the beam should be flush with the top of the joists (fig. 2-115B).

In a concrete block wall, the mason provides the beam notch. When a poured concrete wall is used, a small wood box the size of the notch is fastened to the forms before pouring.

Drop beam.—The drop beam is supported by the masonry piers and the end walls of the foundation. Joists rest directly on the beam (fig. 2-116A). A lap or butt joint can be used for the joists over the beam. In areas of high winds, it is advisable to use a strap anchor or a long bolt to fasten the beam to the piers. One disadvantage of this type of beam is the difference in the amount of wood at the center piers and at the outer walls which can shrink or swell. It is desirable to equalize the amount of wood at both the center and outside walls whenever possible. The flush beam closely approaches this desirable construction feature. The end foundation walls have a notch to provide bearings areas for the ends of the beams (fig. 2-116B). This notch should have the same depth as

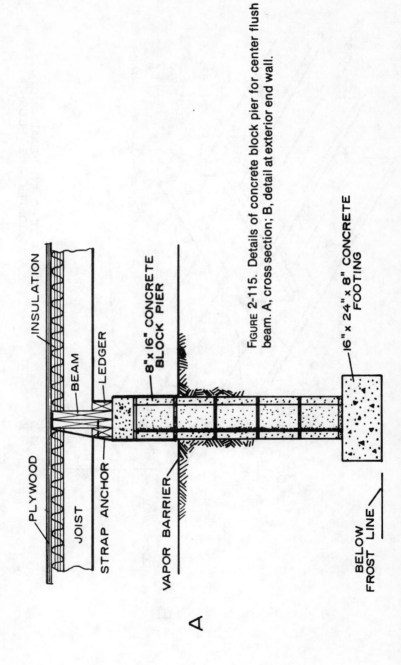

Figure 2-115. Details of concrete block pier for center flush beam. A, cross section; B, detail at exterior end wall.

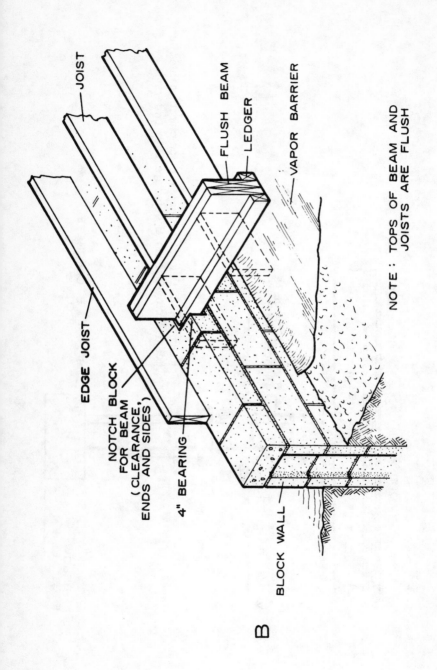

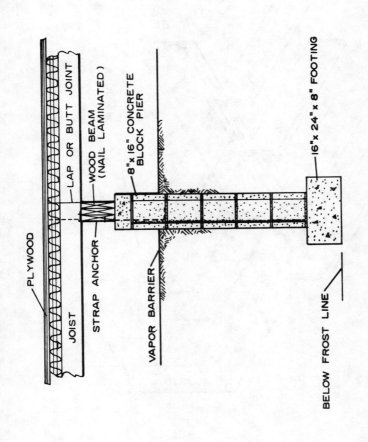

A

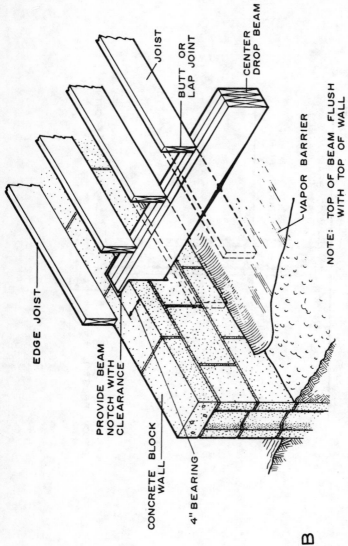

Figure 2-116. Details of block pier for center drop beam: A, cross section; B, detail at exterior end wall.

the beam height. Allow clearance at the sides and end for air circulation. Assembly of the beams, the joist arrangement, and other general details are discussed in the chapter on "Floor Systems" in this manual. Specific details on the size, spacing, and location of the beams and joists are included in the floor framing plan of the working drawings for each house.

Masonry Foundation for Entry Steps

The construction of wood entry steps has been discussed in the chapter on "Steps and Stairs." This type of wood stoop provides a satisfactory entry platform and steps at a reasonable cost. However, when masonry walls are used in the foundation of the main house, it may be desirable to also provide a masonry foundation for the entry steps.

Fig. 2-117A shows the foundation plan for a typical masonry entry platform and step. The walls are normally of 6-inch concrete blocks or poured concrete. Block units 4 inches thick or a 4-inch poured wall have also been used for the supporting front and sidewalls. The size of the top platform for a main entry step should be a minimum of 5 feet wide and 3½ feet deep. The foundation in fig. 2-117A would result in a 6-foot by 3½-foot top with two steps or 6 by about 5 feet when one step is used.

The outer wall and footings are sometimes eliminated in providing support for the concrete steps. Then only the two wing walls are constructed. In such cases, the concrete steps are reinforced with rods to prevent cracking when the soil settles. Use at least two ½-inch diameter rods located about 1 inch above the bottom of the step.

It is important in constructing a masonry entry stoop to tie the wall into the house foundation walls and to have the bottom of the step footings below frost level. A concrete footing should be used for the concrete block wall to establish a level base as well as a bearing for the wall (fig. 2-117B).

The footing should be at least 6 inches wide even though the wall may be less than this. Footings are normally not required for the 6-inch poured wall as the bearing area of the poured wall is usually sufficient. Ties or anchors to the house wall can consist of ½-inch reinforcing rods for the poured wall or standard masonry wall reinforcing for the block wall. These are placed as the house wall is erected. In block construction, both the house foundation wall and the wall for the entry stoop can be erected at the same time and tied together with interlocking blocks.

The finish concrete slab should be reinforced with a wire mesh when fill is used. Concrete is poured over the block wall forming the steps and platform (fig. 2-117B). Boards at each side of the step and platform and one at each riser provide the desired formwork for the concrete.

Skirtboard Materials for Wood Post Foundation

Sheet materials such as exterior grade plywood, hardboard, or asbestos board are most suitable for enclosing a wood post foundation. They need little if any framing and can be installed during or after construction of the house. When the skirtboard will be in contact with or located near the ground, it is most desirable to use a treated plywood when available. However, a 3-minute dip coat of a penta preservative or a water-repellent preservative along the exposed edges will provide some desirable protection for untreated material. It is also desirable to treat the exposed edges of hardboard with water-repellent preservative. Although asbestos board and metal skirtboards do *not* require preservative treatment, they are not as resistant to impacts as plywood and hardboard. Plywood and hardboard can be easily painted or stained to match the color of the house.

Plywood for skirtboards should be an exterior grade to resist weathering. A standard or sheath-

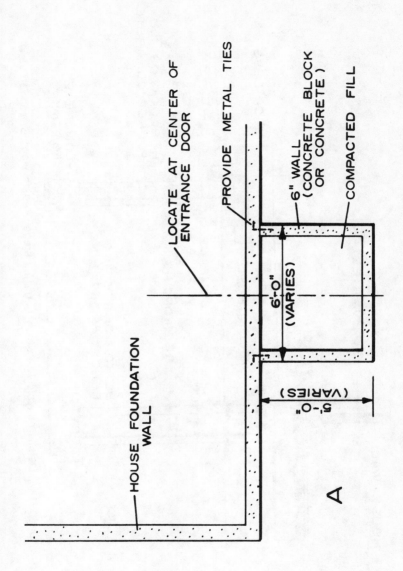

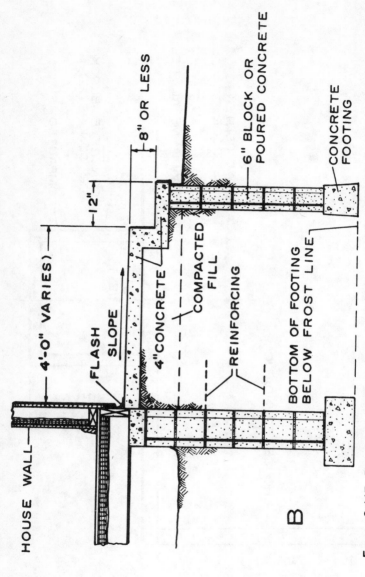

Figure 2-117. Details of typical masonry entrance steps. A, plan view; B, section view.

ing grade with exterior glue is the most economical and can be readily stained. Plywood with rough-sawn or grooved surfaces commonly used for panel siding is also satisfactory.

Tempered hardboard probably provides somewhat better performance than regular-density (Standard) hardboard. However, with paint or similar protective coatings, the lower cost regular-density hardboard should give satisfactory performance. Asbestos board normally requires predrilling to prevent cracking when nailing near the edges.

Skirtboard for House With Edge Beams

Two relatively simple means of fastening the skirtboards to the floor framing can be used. The first method must be employed during construction of the floors and walls. The second can be used either during construction or after the house has been completed.

First method.—Fig. 2-118 illustrates the first method of providing a nailing surface for the skirtboard material. The subfloor and the bottom plate of the walls are extended beyond the edge beam a distance equal to the thickness of the skirtboard (fig. 2-118A). The skirtboard is then nailed to the beam and to the foundation posts. Splices should be made at the post when possible. When splices must be made between the posts, use a 2- by 4-inch vertical nailing cleat on the inside. Fourpenny or fivepenny galvanized siding or similar nails can be used for the ¼-inch hardboard or asbestos board. Space them about 8 inches apart in a staggered pattern. Use sixpenny nails for ⅜-inch plywood and sevenpeny or eightpenny nails for ½-inch plywood skirtboards with the same 8-inch spacing. The panel siding is nailed directly over the top portion of the skirtboard. Nailing recommendations for the siding are given in the chapter on "Exterior Wall Coverings."

The detail for the end wall in this type of installation is shown in fig. 2-118B. A 2- by 4-inch cleat should also be used for the skirtboard joints when they occur between the foundation posts.

Some support or a backing may be required for the bottoms of the skirtboard to provide stability and resistance to impacts from the outside especially with thinner materials. Treated posts, treated 2- by 4-inch stakes, or embedded concrete blocks can be used for this purpose. Treated posts or stakes can be driven behind the skirtboard and the concrete blocks can be embedded slightly for added resistance (fig. 2-118 A and B). Space these supports about 4 feet apart or closer if required.

Second method.—A second method of installing the skirtboard can be used either after the house has been completed or during its construction. This system consists of nailing the skirtboard to the inner face of the ledger (fig. 2-119). The skirtboard must be fitted between the posts which are partly exposed. Use the same method of nailing and blocking at the bottom of the boards as previously described. In addition, toenail the ends of the boards into the posts which they abut. Fig. 2-119 A shows the details of installation at a sidewall and fig. 2-119 B shows the details at an end wall. Use treated posts or stakes or concrete blocks as backers at the bottom of the skirtboard as previously described.

Skirtboard for House With Interior Beam

In those crawl-space houses with interior supporting beams and posts and with side overhang, the application of skirtboards is much the same as when the beams are located under the outside walls. The skirtboard may be nailed to the outside of the joist header or to the inner face (fig. 2-120 A). When it is fastened to the exterior face, the subfloor and the bottom wall plate are extended beyond the header a distance equal to the thickness of the skirtboard material. The skirtboard can also be fastened to the inside face of the header, but then it must be notched at each joist (fig. 2-120 A). Use the same type of nails and nailing patterns described for the details shown in fig. 2-118 and 2-119.

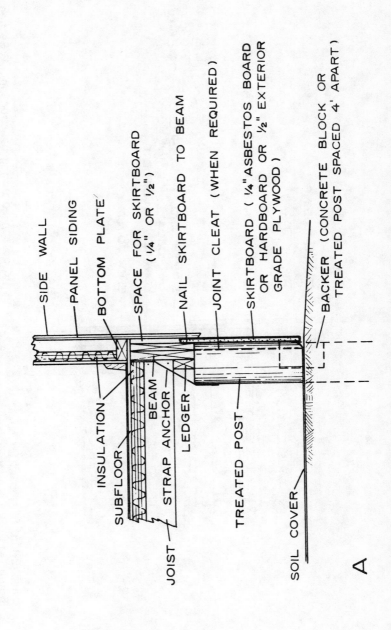

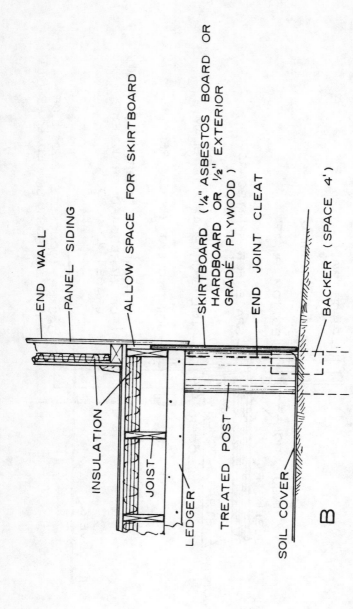

FIGURE 2-118. Skirtboard for house with edge beams (to be applied during construction). A, Section through sidewall; B, section through end wall.

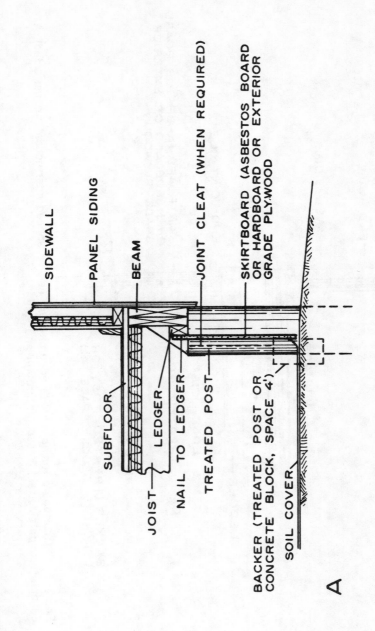

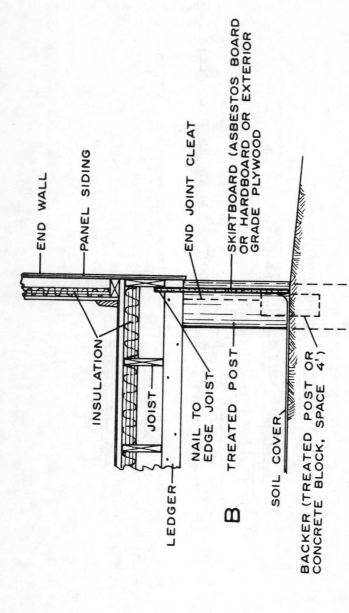

Figure 2-119. Skirtboard for house with edge beams (can be applied after house is constructed). A, Section through sidewall; B, section through end wall.

Details at the end walls of the house are much the same as those at the sidewalls (fig. 2-120B). When the skirtboard is nailed to the inner surface of the edge joist, no notching is required except at the center and edge beams. Use backers for the bottom of the skirtboard as previously described.

Soil Cover, Ventilation, and Access Door

An enclosed crawl space should not only be protected with a soil cover but should also have a small amount of ventilation. A soil cover of 4-mil polyethylene or similar vapor barrier material placed over the earth in the crawl space will minimize soil moisture movement to floor members and floor insulation. Lap the material 4 to 6 inches at the seams, and carry it up the walls or skirtboards a short distance. Use sections of concrete block or small field stones to keep the material in place.

When the use of a soil poison is required in termite areas, do not install the soil cover for several days after poisoning or until the soil becomes dry again.

Ventilators should be installed on two opposite walls when practical. Standard 16- by 8-inch foundation vents can be used in concrete block foundations. Usually, in small houses, two screened ventilators, each with a net opening of 30 to 40 square inches, are sufficient when a soil cover is used. Crawl spaces with skirtboard enclosures should also be ventilated with small screened vents.

Access doors should be provided in crawl spaces. With a masonry foundation, they can consist of a 16- by 24-inch or larger frame with a plywood or other removable or hinged panel. Install the frame as the wall is constructed. With skirtboard enclosures, provide a simple removable section. When practical, the access doors should be located at the rear or side of the foundation and at the lowest elevation when a slope is present.

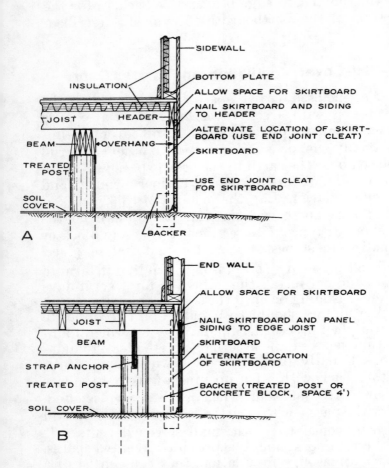

FIGURE 2-120. Skirtboard for house with interior beam (with overhang). A, Section through sidewall; B, section through end wall.

3

Designs for Low-Cost Wood Homes

For over half a century the Forest Service, U.S. Department of Agriculture, has pioneered in research to more effectively utilize wood and wood products. In an effort to help alleviate the Nation's housing crisis for low-income families, the Forest Service has applied this research expertise to the design of low-cost wood homes.

Eleven house designs have been developed of varying style and size, which can be built for approximately one half of normal construction costs. These homes have all the essentials necessary to provide comfortable living for families with up to 12 children. They are intended primarily for rural America where housing for low-income families is often well below acceptable standards.

Economies have been made through simplicity of design and elimination of frills, by specification of economical but durable wood materials, and by employing unconventional new materials, systems, and uses of wood and wood products. The fact that these homes are "low-cost" does not mean that they use second-rate materials or construction methods. Neither strength, safety nor durability have been sacrificed to obtain reduced costs.

Some of the designs are for homes of a conventional appearance, some are for homes of a more innovative nature. None are meant to represent luxury living. They all have one aim: to obtain as much comfortable living space as possible for people of very limited means.

The Forest Service does not build or market these homes. But it is the hope of the Service that others will make use of the designs in helping to meet the need for low-cost housing.

This publication briefly describes the eleven designs. Five of the designs were developed at the Forest Products Laboratory at Madison, Wis., by L.O. Anderson. Six of the designs, including the tubular, duplex, and round houses, were developed at the Southeastern Forest Experiment Station in Athens, Ga., by H.F. Zornig.

Working plans for construction of these homes can be obtained from the Superintendent of Documents at Washington, D.C. fo a nominal fee.

House Plan

This house was developed to provide a good livable home for a cost much lower than most houses now being constructed. It is 24 by 32 feet in size and contains 768 square feet of living area. In spite of its relatively small size, this home has three bedrooms, affording desirable privacy for a family with 3 to 5 children. There is little waste space and the bath and kitchen, as well as the living room and dining area, are conveniently arranged.

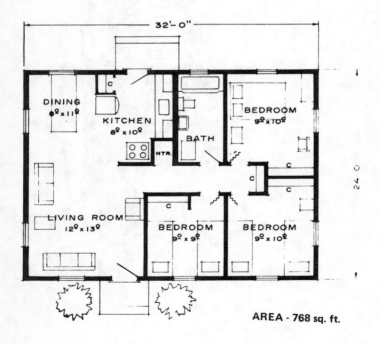

AREA - 768 sq. ft.

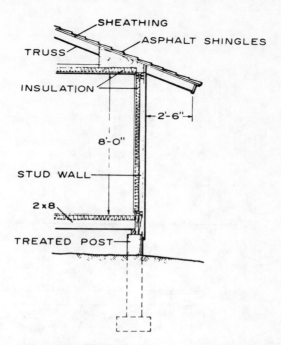

This crawl-space house has a treated wood post foundation which reduces the cost without reducing its performance. It can be constructed on sloping sites without costly grading and masonry work. Most of the materials used can be obtained at local lumber yards or small local mills.

The panel siding exterior and trim are finished with long-lasting stains which can be obtained in many contrasting colors. The wide overhangs at the cornice and gable ends provide a good appearance as well as excellent protection for the side walls. Insulation in the walls, floors, and ceiling reduce heating costs as well as providing a cool house during the hot summer months. The open living-dining-kitchen area gives a feeling of spaciousness not possible when walls separate these rooms.

Designed By L. O. Anderson
Forest Products Laboratory

DINING ROOM

House Plan

This home was developed for a large family of up to 12 children at a reasonable cost. It is 24 by 36 feet in size and is one and one-half stories. The first floor has 864 square feet, consisting of three bedrooms, a bath, and a living-dining-kitchen area. The second floor contains about 540 square feet and consists of two large dormitory-type bedrooms. Each is divided by a wardrobe-type closet which, in effect, contains space for two single beds on each side.

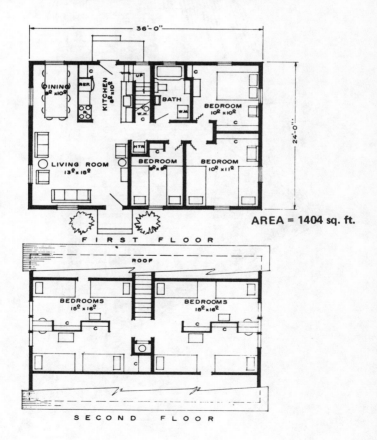

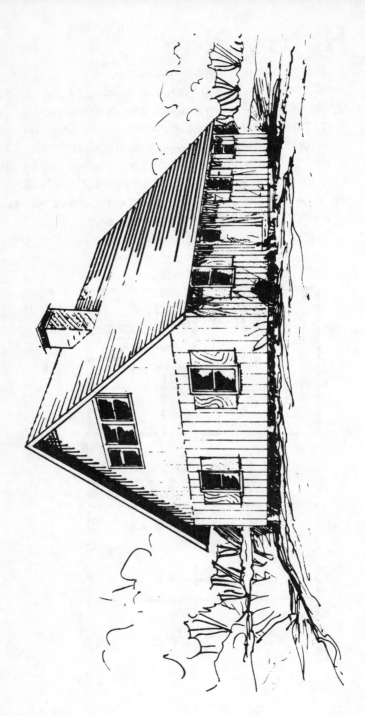

The design features which aid in reducing the cost and in providing maximum space for the overall size are (a) the treated post foundation with a crawl space and (b) the steeply sloped roof. The long-lived treated foundation posts can be installed with little or no costly grading and leveling. They serve as a rugged base for the beams and joists of the floor system. The one-half pitch (12 in 12 slope) roof with a 4-foot-high knee wall encloses two large 15- by 16-foot dormitory bedrooms. Windows are installed at the gable ends of the second floor.

The panel siding on exterior walls and the trim and shutters are finished with a pigmented stain which can be obtained in a variety of colors for contrast. The wide overhangs at both the gable ends and the cornice provide desirable protection to the side and end walls. Floors, walls, and ceiling are insulated to reduce heat loss. The thickness of this insulation can be varied, and the amount usually depends on whether the house is located in the northern part of the United States or in a milder climate.

Designed By L. O. Anderson
Forest Products Laboratory

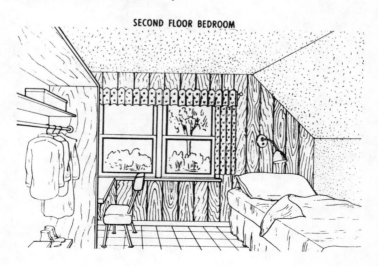

SECOND FLOOR BEDROOM

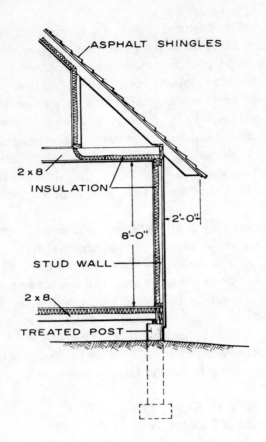

House Plan

This house plan was developed to serve a small family or senior citizens who require only one main bedroom and a spare room which can serve as a sewing room, workroom, or also as a second bedroom. The kitchen and living room are open with a drop beam between them. The main house is 24 by 24 feet in size and has 576 square feet of living area. As shown in the plan, an 8- by 12-foot porch is included as well as a carport. When a minimum house is desired, the porch and the carport can be eliminated.

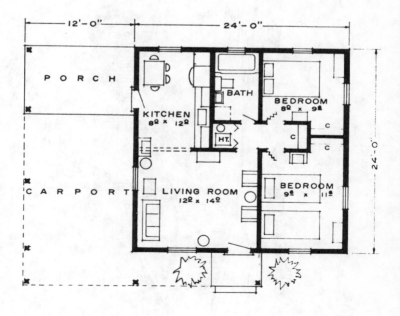

AREA - 576 sq. ft.

This crawl space house has a treated wood post foundation which reduces the cost of the house without reducing its performance. This type of house can be constructed on a sloping lot, eliminating costly grading which is normally necessary for houses with other designs.

The floors, walls, and ceilings are well insulated, insuring comfort both winter and summer. The small forced-air heating unit can be either the oil or the gas-fired type. The bathroom contains space for a small washing machine. Closet space is sufficient for a small family. More storage area can be provided by the installation of small plywood wardrobes in the bedrooms.

The panel siding and other exterior wood can be finished with a pigmented stain. These materials are not only easily applied, but they are long lasting. Many colors are available. The interior has a gypsum board wall and ceiling finish which are painted. In addition, the bath and part of the living room walls have prefinished plywood for a desirable contrast.

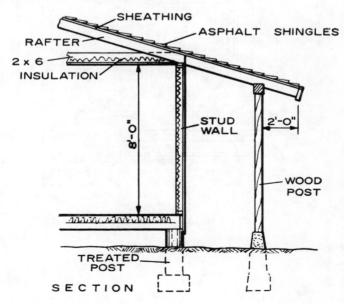

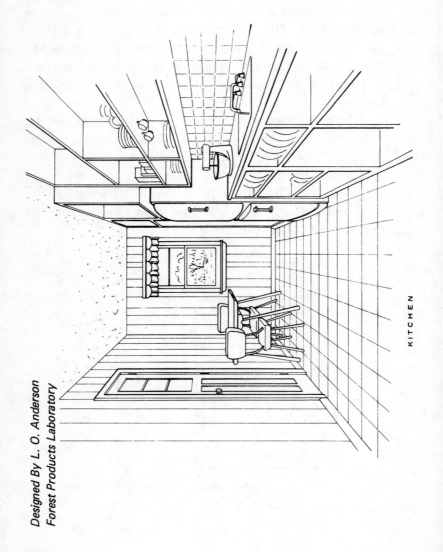

*Designed By L. O. Anderson
Forest Products Laboratory*

KITCHEN

House Plan

This home is an expandable type. With its steeply pitched roof, there is more than adequate space on the second floor for two dormitory-type bedrooms, which can accommodate up to eight children. The working drawings also provide for an

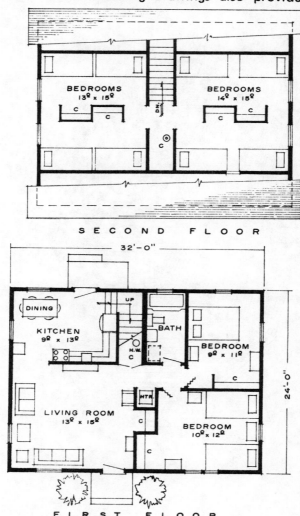

AREA = 768 sq. ft. expandable to 1228 sq. ft.

additional bath on the second floor if it is desired. The house is 24 by 32 feet in size with an area of 768 square feet on the first floor and about 460 square feet of usable space on the second. The first floor contains a moderate-size living room, a compact kitchen with a large adjoining dining area, two bedrooms, and a bath. Storage space is adequate with four closets on the first floor and five on the second. The first-floor bath is arranged to accommodate a washing machine.

There are several important factors which aid in reducing the cost of this home. One is the fact that it is a crawl-space house, which eliminates the need for extensive excavation and grading. In addition, the floor framing is supported by long-lived treated wood foundation posts resting on concrete footings. A more costly masonry foundation is included in the working drawings as an alternate. The use of a single covering material for the subfloor and the exterior walls also leads to reduced costs. The subfloor consists of tongued-and-grooved plywood or square-edge plywood with edge blocking and serves as a base for a resilient floor covering. The panel siding, with perimeter nailing, eliminates the need for corner bracing as well as the need for sheathing. Such coverings are usually rough-textured exterior-grade plywood, which can be finished with a pigmented stain. Suitable stains are available in many colors, and contrasts can be obtained by treating the trim and shutters with a different color or shade.

Further cost reductions are obtained by eliminating much of the exterior trim as well as some of the less important interior millwork. However, these refinements can be added in the future. An adequate forced-air heating unit with relatively short heat runs is also a part of the design. Insulation is included in the floor, wall, and ceiling areas; the thickness selected depending on the location of the home. In the colder climates, the ceiling and floor insulation might be 4 inches or thicker with 2-inch-thick blanket insulation in the walls.

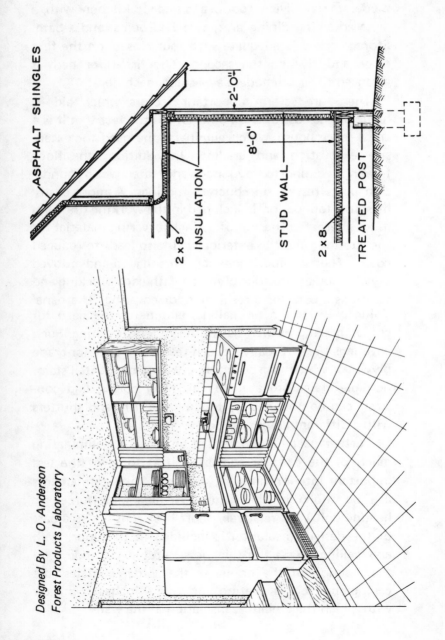

*Designed By L. O. Anderson
Forest Products Laboratory*

House Plan

This plan provides for the construction of either a single-story two-bedroom home or an expandable type with two additional bedrooms on the second floor. The basic house is 24 by 28 feet in size with an area of 672 square feet. The expandable plan provides an additional area of about 370 square feet on the second floor. The second-floor bedrooms can be completed with the rest of the house or left unfinished until later. Both plans include a kitchen, bath, living room, and two bedrooms on the first floor.

To accommodate the stairway to the second-floor bedrooms in the expandable plan, the first-floor bed-

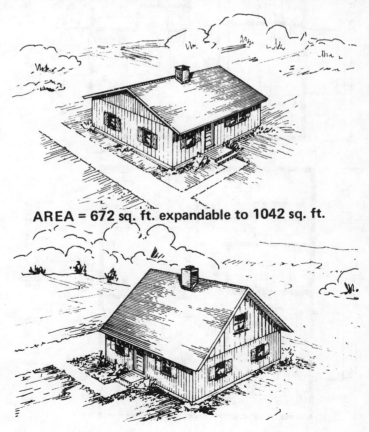

AREA = 672 sq. ft. expandable to 1042 sq. ft.

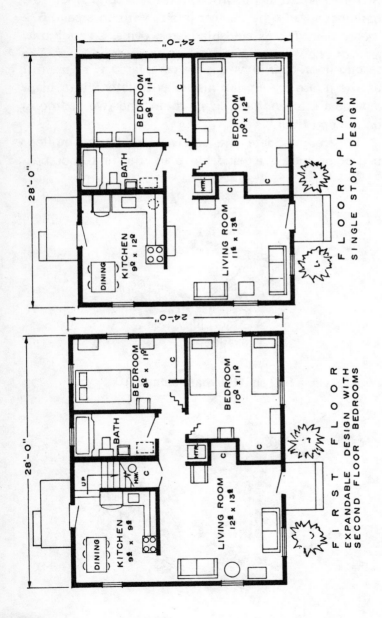

rooms and the kitchen are slightly smaller than those in the one-story plan. One second-floor bedroom is 9-1/2 by 14 feet and the other is 13 by 14 feet. The larger bedroom may be divided by the addition of a wardrobe wall, which provides two closets and also serves as a room divider.

Both plans have a front entrance closet and a closet for each bedroom. The expandable house also has a storage area under the stairway in which the hot water heater is located. The heating unit is located in a small closet adjacent to the bath-bedroom hallway. Walls, floors, and ceiling areas are insulated. Dining space is provided for in each kitchen.

There are a number of factors which aid in reducing the cost of these homes. They are designed as crawl-space houses with post or pier foundations, which eliminate the need for extensive grading on sloping building sites. The single floor covering serves as a base for resilient tile or a low-cost linoleum rug. It can be painted if further cost reductions are required. Panel siding is used for exterior wall finish which eliminates sheathing and the need for a braced wall. Exteriors are finished with long-lasting pigmented stains. Many contrasting colors are available in this type of finish. Exterior and interior trim and millwork have been reduced to a minimum. However, many of these refinements can be made at anytime after the house has been completed.

Details of a second-floor bathroom, a porch addition, a full foundation wall, and an enclosing skirtboard are also included in the working drawings.

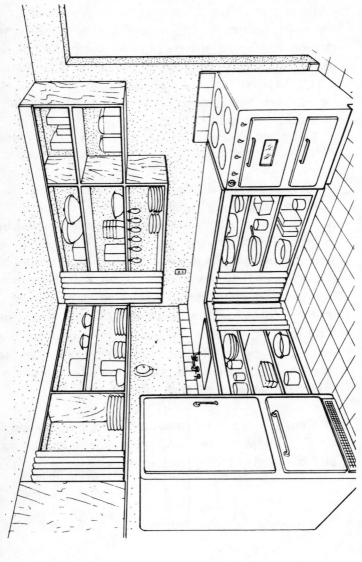

Designed By L. O. Anderson
Forest Products Laboratory

House Plan

This design, intended for a flat site, encloses 1,024 square feet. The square plan provides much more usable space, within the same exterior walls, than a typical rectangular plan.

While essentially conventional, the design features a novel floating-wood-floor system. The space under the floor serves as a return-air plenum from a centrally located forced-air furnace. Air will flow into each room through the opening above the door, then enter a space behind the baseboard for return under the floor to the furnace. Alternate designs are available for a conventional wood floor over a crawl space and for a concrete slab floor.

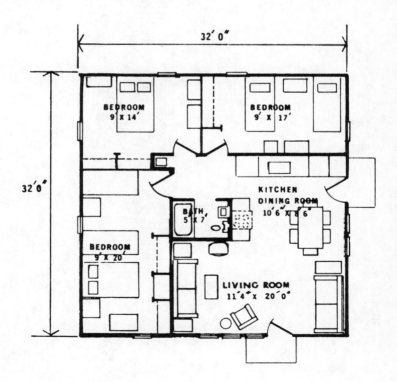

AREA = 1024 sq. ft.

Large bedrooms, closets, built-in desks, and shelving in the bedrooms are bonus features. These bedrooms are really multi-purpose rooms for play, study and sleeping.

Economical, nonbearing partition walls of particle board have replaced conventional hollow-core interior walls. Being nonbearing, they can be located in various positions without affecting the structure.

Plywood combination siding and sheathing on the exterior walls can be finished with modern natural finishes and stains for durability and easy maintenance.

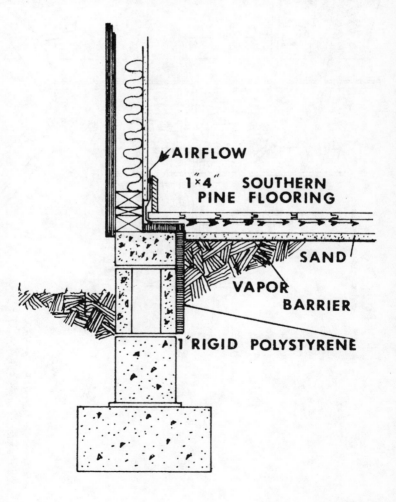

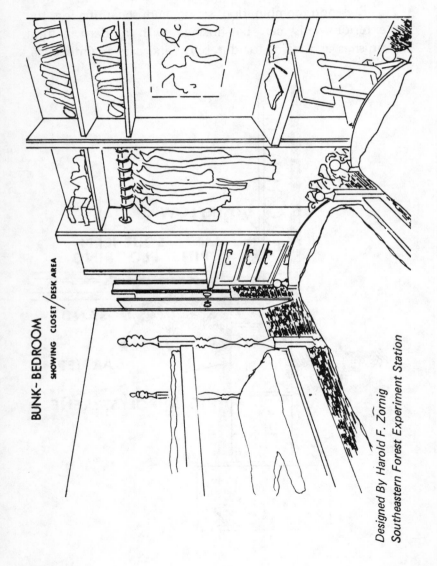

BUNK-BEDROOM SHOWING CLOSET/DESK AREA

Designed By Harold F. Zornig
Southeastern Forest Experiment Station

House Plan

Wood piers are used for the foundation to make this plan particularly suited to sloping sites. The piers are pressure treated with a clean, non-leaching preservative. A carport and storage area under the house might be considered for steeply sloping sites.

The play area is a bonus feature in a three-bedroom house of this size. Elimination of a central hall has provided more usable living space in the 1,008 square feet of floor area. The large bedrooms will also provide extra, multi-purpose space for large families.

Conventional framing, stock-sized windows and doors, and the simple rectangular plan were selected with the small contractor in mind. Rafters and joists, instead of roof trusses, help to open up the attic space for convenient storage. Plywood combination sheathing and siding on exterior walls is finished with a natural finish or stain to provide atrractive, economical, easy-to-

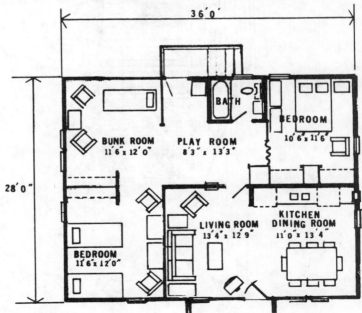

AREA = 1008 sq. ft.

maintain exterior walls. The skirting could be placed around the foundation piers in cold climates to lower the heating bill. If this is done, the floor insulation and insulation board under the floor joists could be eliminated.

Heat is provided by a centrally located forced-air wall furnace, or by electric heaters.

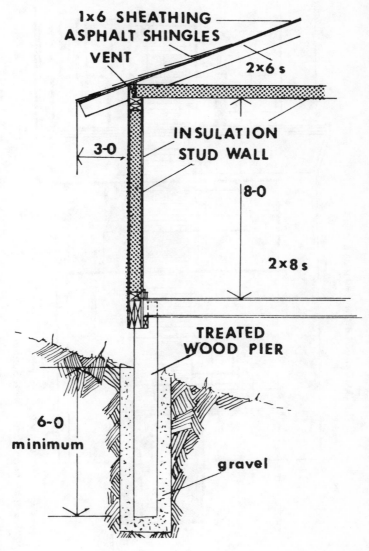

PLAYROOM and BATH — UTILITY AREA

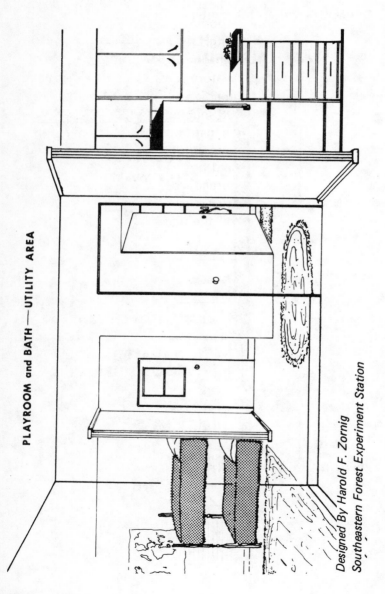

Designed By Harold F. Zornig
Southeastern Forest Experiment Station

404

House Plan

Strong, durable, pressure-treated wood poles support this house, which is particularly suitable for sloping sites.

This attractive house was designed expecially for large families wanting a low-cost three-bedroom home. The bedrooms and dining room will accommodate a large family without crowding. A utility room has been provided next to the bathroom, and the laundry tub will be useful as a second lavatory. The sizable open space of the kitchen-dining-living area will make this house seem much larger than its 1008 square feet.

When this house is constructed on a sloping site, a carport and storage room under the house are bonus features. The furnace and hot water heater will usually be located in the storage room under the house; however, if the site is level, the furnace could be in the linen closet location, and the under-counter type of hot

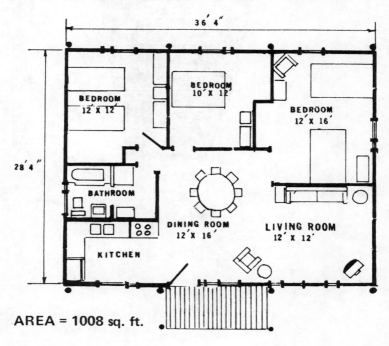

water heater could be installed in the kitchen.

This pole-truss structural system has several advantages over conventional foundation and framing systems. It is economical, very little site grading will be needed, and construction need not be delayed by weather or frozen ground. In addition, all walls are non-load bearing. This makes all walls easy to prefabricate and erect, and it provides design flexibility in that the interior walls can be easily moved to alternate locations.

Since all of the walls, except those around the plumbing core, are prefabricated in 4 x 8 foot panels, window and door panels have also been designed for use with the wall panels. This prefabricated panel system of construction can be used very effectively with pole-frame construction.

Southern pine flooring is used throughout the house. A single layer of tongue and groove, 1 x 4 inch strip flooring is fastened to the joists with nails and adhesive. This provides a strong, stiff, and durable floor. Inlaid linoleum protects the floor in the bath and utility area.

Plywood combination siding and sheathing, finished with a natural finish or stain, provides attractive exterior walls, that are economical and easy to maintain.

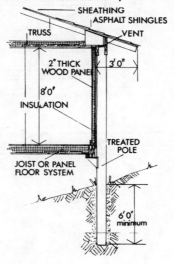

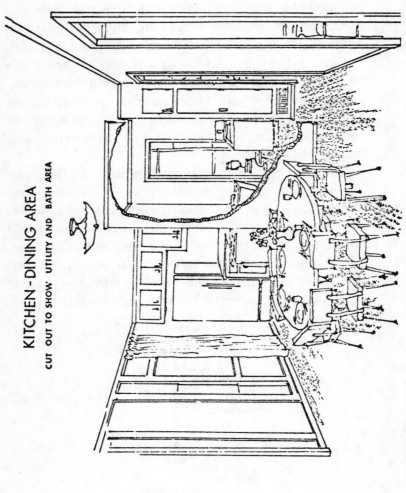

Designed By Harold F. Zornig

House Plan

This unusual home offers attractive living space within its curved walls. It is primarily intended for sloping sites in rural areas. It also has particular advantages for second homes on lakeside or ocean front sites. Very little site disturbance is necessary for construction. This home provides a total of 1,000 square feet of floor area, with either two or three bedrooms on the second floor. The estimated cost of this home is about one-half the cost of equivalent conventional construction. Lumber and plywood requirements are also about half of such conventional construction.

This design is one of several produced under a research program currently underway at the Southeastern Forest Experiment Station, Forest Service, U.S. Department of Agriculture. The objective of this program is to develop designs and techniques that will lead to the more efficient use of wood and wood products in housing for low-income families in both rural and urban areas.

Foundation System

Eight preservative-treated 6x6 square posts support this home. Treated round poles could be used instead. The posts extend up to the second floor and are bolted to the laminated ribs and the floor beams. Embedment and footing requirements will depend on local soil conditions and the height of the house above the ground.

Exterior Curved Walls

Eight egg-shaped, glue-laminated wood ribs form the curved exterior walls. The ribs are manufactured in half sections. They should be produced by an experienced

AREA = 1000 sq. ft.

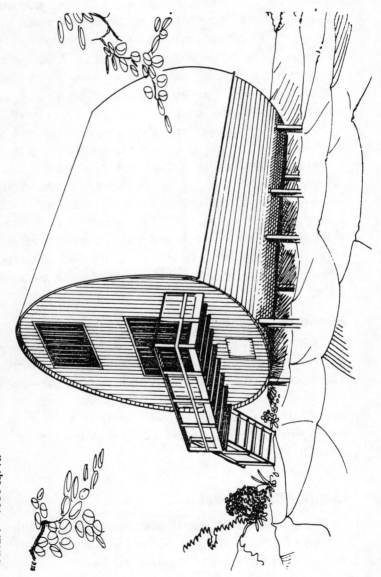

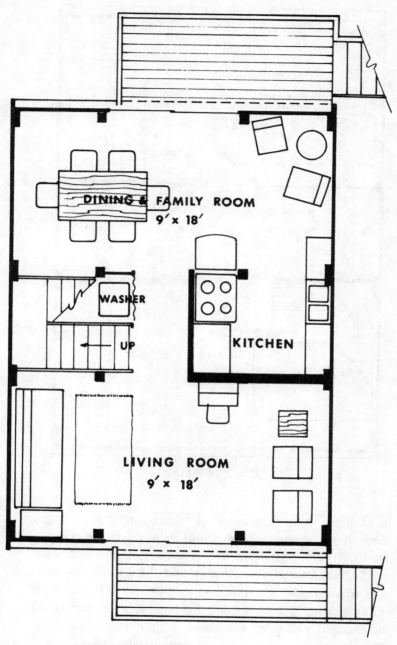

FIRST FLOOR

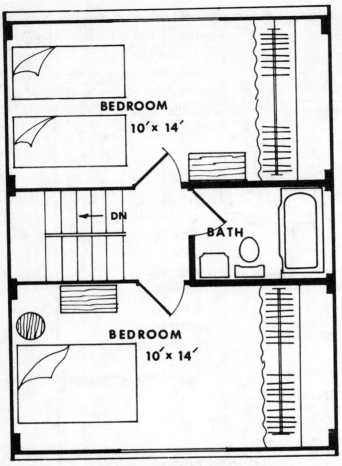

SECOND FLOOR

laminator. A complete rib is located at the front and back faces of each pair of foundation posts. The ribs are bolted to the posts and to the ends of each floor beam.

Tongue-and-grooved 2x6 wood decking is nailed to the ribs around the entire circumference of the house. The decking is covered on the outside with foamed-in-place rigid polyurethane down to the first floor level. The foam provides excellent thermal insulation and a water-tight roof covering. It is painted on the outside with one or two coats of aluminum-filled asphalt emulsion, or a roof paint of similar quality. Below the first floor, the foam is placed on the inside of the decking to insulate and seal the crawl-space. If a forced-air furnace is used, the sealed crawl-space can serve as an air plenum. The crawl-space also provides a large storage area, accessible from the outside.

Exterior End Walls

Exterior 3/8-inch plywood is placed over 2x4 framing to enclose the end walls. These walls will resist racking loads and stiffen the entire structure. Interior finish is 3/8-inch gypsum board, or 1x6 wood paneling placed horizontally. Conventional insulation is placed in the walls. The frame members are installed flatwise so that they can be face nailed to the ribs and floor joists. A cantilevered deck is provided at each end of the home by extending the first-floor joists through the end walls. Joists on the second floor can also be extended for an upper deck if desired.

Floor System

Single layer wood-strip flooring is placed directly on the floor joists over a 1/4-inch bead of construction adhesive and nailed at each joist location. Construction adhesive will insure a stiffer floor and help to eliminate squeaking. The joists are exposed, and the lower side of the flooring on the second floor is the ceiling of the first floor.

Interior Partitions

Single-layer, self-supporting, panels of particleboard or plywood are used to partition interior spaces.

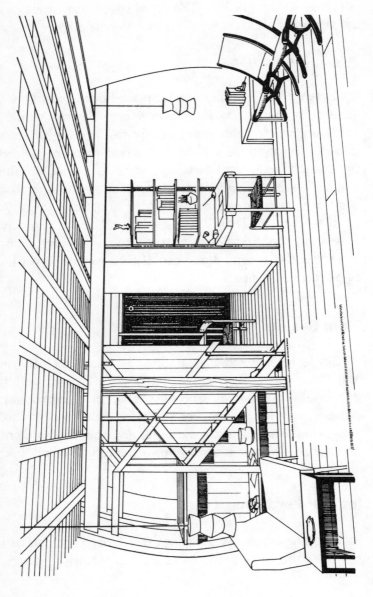

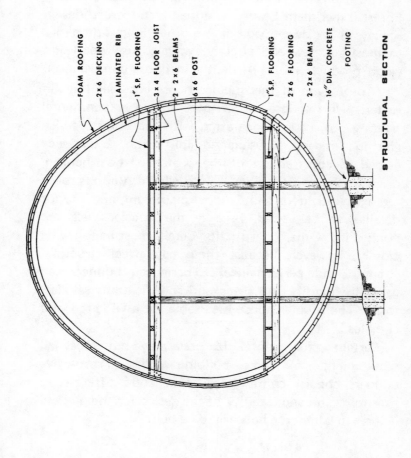

Tongue-and-grooved solid wood paneling can also be used. Partitions are installed without framing except where a hollow wall is needed to enclose wiring, pipes, or ducts.

Additional Information

The Housing Research Unit of the Southeastern Forest Experiment Station is not an architectural design agency. It has developed this and other designs primarily to illustrate new and effective ways to use wood and wood products more efficiently in house construction.

A prototype of this tubular house has been built in eastern North Carolina as part of an experimental housing project in cooperation with the Forest Service and the Federal Housing Administration. Experience gained in construction of the prototype was incorporated in the revised plans now available. Various room arrangements, door and window placement, and interior details can be varied, because the interior walls are nonbearing. Other construction details must be followed closely, however, because they may affect structural strength and performance. Experienced builders can generally modify the plans and specifications satisfactorily. The tubular house is suitable for a wide range of climates.

Certain experimental features may not meet all requirements of some building codes. Prospective builders should confer with local code officials to determine the applicability of the design for the particular area in which the house is to be built.

Designed By Harold F. Zornig
Southeastern Forest Experiment Station

House Plan

This unique design provides a three-bedroom home with 1,134 square feet of living area. It is designed for a flat site. A smaller version provides three bedrooms and a total area of 804 square feet.

General Background

Round homes are an efficient means of providing housing space. In this design, interior walls are spaced radially from a central atrium hall. The design permits good arrangements of rooms and furniture. It includes a number of experimental features that are being evaluated in the laboratory and in full-size prototypes built under the auspices of the Experimental Housing Program of the Federal Housing Administration.

It is estimated the round house will cost about half as much as a conventional house with an equal amount of floor space. The house should be easy to build and suitable for self-help programs.

Floor System

A circular concrete slab is placed within a low brick foundation wall. The perimeter insulation and vapor barrier are conventional. Where needed, the soil is treated or poisoned to prevent attack by termites and other insects. A preservative-treated wood member is fixed around the edge of the slab. The ends of the exterior plank walls are nailed to this member.

Partition Walls

The interior partition walls are particleboard panels located under the roof beams and fitted into slots in 2x2

AREA = 1134 sq. ft.

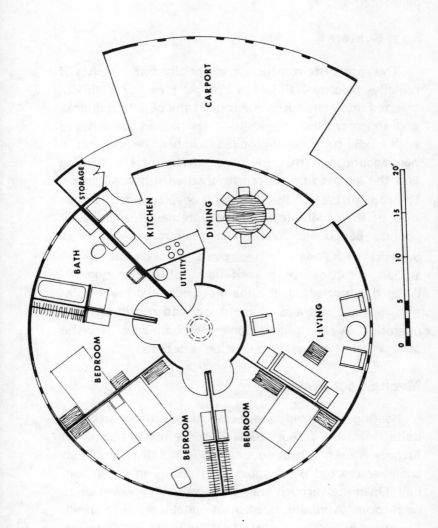

members at the vertical joints. The panels can be moved to alternate locations to provide two, three, or four bedrooms. Closet shelving and sidewalls serve to stiffen the particleboard walls. The walls may be finished with conventional wall paints or with natural finishes, either pigmented or clear.

Roof System

The roof system, which is essentially flat, consists of radially placed 4x6 beams, rough sawn or finished, inserted into slots cut in the tops of the perimeter planks and those around the atrium. The beams are covered with 1x6 tongue-and-grooved lumber decking in a herringbone pattern. The lower surface of the decking and the exposed beams are finished with natural stains. The roof surface of the decking is covered with a 1-inch layer of foamed-in-place rigid polyurethane insulation (2 pounds per cubic foot). The polyurethane is then overlaid with one or two coats of aluminum-filled asphalt emulsion, or a roof paint of similar quality. When thus protected, the foam provides thermal insulation, a good seal against moisture, and roofing in one operation. A clear plastic dome may be placed over the atrium hall, if natural illumination is desired.

Mechanical Systems

Heating is provided either with electric baseboard units, or with a furnace located in the utility room. If a furnace is used, a heat duct carries warm air to a plenum chamber created by dropping the ceiling in the atrium hall. Openings through the plank wall carry warm air to each room. A minimum of sheet metal work is required. Electrical outlets around the exterior walls and along the interior partitions are provided in a wood baseboard-raceway system. A water heater is installed in the utility

room. Some of these features are illustrated in the sketch of the atrium wall section.

Exterior Wall System

Exterior walls consist of rough-sawn 2x8 softwood planks, placed on end around the slab and joined with hardboard splines inserted in slots in the plank edges. To minimize dimensional changes in service, moisture content of the planks should be no more than 12 to 16 percent when they are installed. A fascia board is fitted around the edge of the roof. The wall of the atrium hall is of similar plank construction. Exterior wall surfaces are finished with a pigmented natural finish containing a preservative and water repellent. This attractive finish is easy to apply and maintain and it has good durability. Interior wall surfaces are finished with pigmented or natural stains.

Window System

Sections of two adjacent planks are omitted at appropriate locations in the exterior wall, and fixed glass panels are fitted into the openings, framed with a simple wood frame, and properly sealed. Ventilation is provided by a system of vertically sliding hardboard panels. As an alternate method, single-hung aluminum windows are fitted into the plank openings.

Carport

The extended roof over the carport is supported by another plank wall, with appropriate openings. The roofline is also extended over exterior doors by cantilevering interior roof beams out beyond the exterior walls.

Additional Information

Two prototype homes of this design have been built by a commercial contractor and are now occupied in Eastern North Carolina. Experience gained in construction of these homes was incorporated in the revised plans now available. Various room arrangements, door and window placement, and interior details can be varied, because the interior walls are all nonbearing. Other construction details must be followed closely, however, because they affect structural strength and performance. Experienced builders can generally modify the plans and specifications satisfactorily. Although the design was intended primarily for warmer southern climates, some modification and insulation of the exterior walls should make the house suitable for northern climates.

Certain experimental features may not meet all requirements of some building codes. Prospective builders should confer with local code officials to determine the applicability of the design for the particular area in which the house is to be built.

Designed By Harold F. Zornig
Southeastern Forest Experiment Station

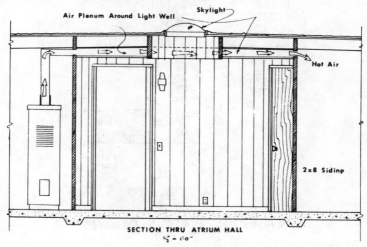

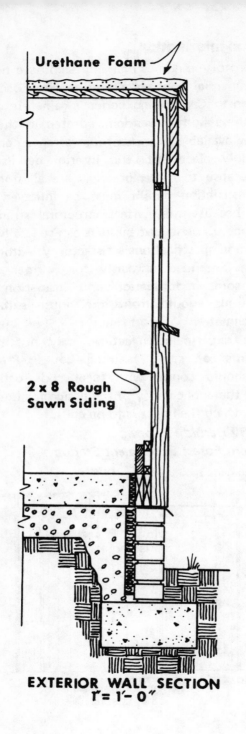

House Plan

A HILLSIDE DUPLEX OF WOOD

This interesting design for a two-family home is intended particularly for sloping sites. It provides a total of 900 square feet in each of the two units, approximately half on each of two floors. The design is based on a pole-frame structure combined with wood arches that can be built in a simple shop. The compact design gives surprisingly open space, with complete privacy for each family. A pleasing wood deck is provided. One unit can be built separately for single-family homes. The simple design makes it particularly attractive for second homes to be built in isolated areas. Estimated costs are about one half that of conventional construction of the same size. Lumber and plywood requirements are about half of such conventional construction.

Foundation System

Preservative-treated poles or posts are set into the ground with a minimum of soil disturbance. Posts extend to the second-floor level. Wood girders are bolted to the posts for each floor, and a conventional wood-joist system is installed. One layer of softwood tongue-and-grooved flooring is nailed to the joists, with a bead of construction mastic adhesive applied to the top of joists for increased stiffness and to reduce floor squeaking. No subfloor is required. Joist spaces of the first floor are insulated, and the lower joist surfaces are covered with plywood or other suitable sheet material.

Roof System

The unique feature of the design is a simple Gothic-type arch, made in two sections, as illustrated on pages 3 and 4. An arch consists of 1x2 lumber flanges nailed to short length of 2x4's with spacing between them. Arch

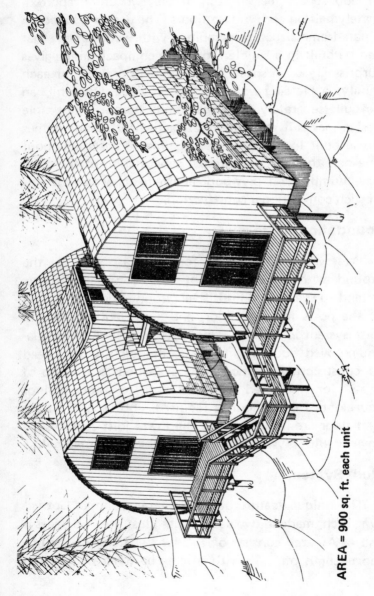

AREA = 900 sq. ft. each unit

sections can be assembled on a simple jig. Each section weighs only about 30 pounds. The sections are assembled on the ground, and the joist for the second floor is nailed to the arch. The sections are joined at the peak with a plywood gusset. The assembly is then raised into place and fastened to the ends of the first-floor joist.

Several roof sheathings may be used. In one prototype, erected in Athens, Georgia, 1/2-inch softwood plywood was used, with arches 24 inches on center. The plywood was covered with building paper and asphalt shingles with a stapled lower tab so that shingles were tight on the reverse curvature of the lower roof sections. Alternate sheathing might be bevelled siding nailed to the arches and finished with a water-repellent pigmented stain. The latter system is easier to install without scaffolding, and the problem of bending plywood in large sheets to the slight curvature of the arches is avoided. Insulation is installed between arches, and the inner surface is covered with thin gypsum board, plywood, or lap siding to provide various interior treatments.

End Walls and PartitionsSpiral-Wood Stairs

All walls are essentially nonbearing, although end walls do provide stiffening of the arch frame. End walls are 2-inch thick lumber-framed panels, erected in place, covered with plywood on the outside, and finished with natural or pigmented stains. The inner surface may be covered with plywood or other panel materials. Insulation is provided in the spaces between framing. Aluminum or wood windows are installed as desired. Sliding patio doors in each unit open to the cantilevered deck.

Interior partitions may be of a single thickness of particleboard with panels fitted into slots in vertical 2x2 member at joints. The walls can also be of conventional construction.

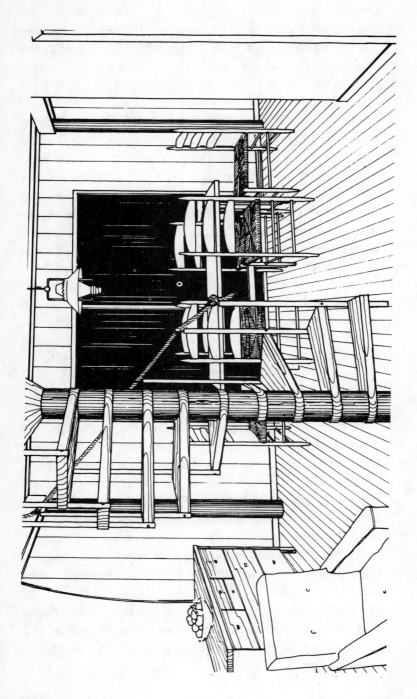

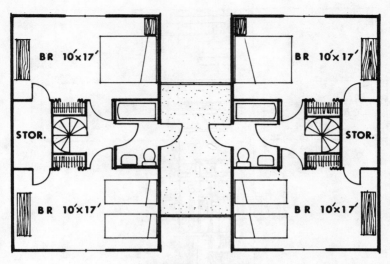

SECOND FLOOR

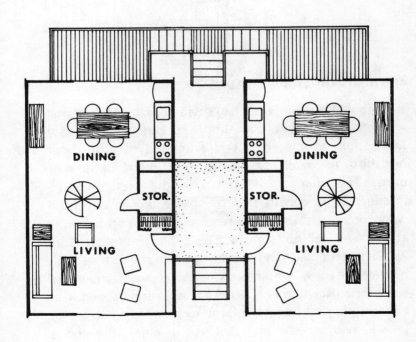

FIRST FLOOR

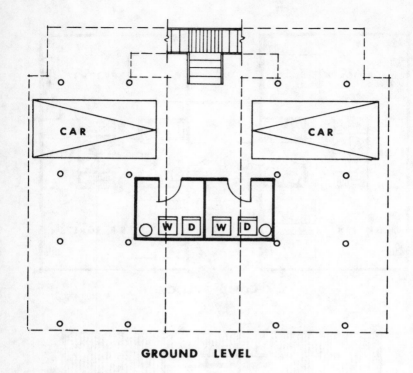

GROUND LEVEL

Spiral-Wood Stairs

Although the units can be provided with conventional wood stairs, a spiral-stair unit was designed for easy fabrication on the site. It is estimated to cost about one-third as much as conventional metal spiral-stair units. The stairway is shown in the sketch of the interior. It is constructed of spacer blocks cut from wood poles, and drilled to receive a metal rod threaded at the ends for bolts and washers. Treads are cut from 2x12 timbers, and sandwiched between the spacers with mastic adhesive. Metal pipe balusters are fastened to edges of adjacent treads, front to back, and capped with a wood plug into which a large screw eye is inserted to take a rope banister or rail. Once in place, the unit is tightened with the bolts at top and bottom. The wood can then be stained.

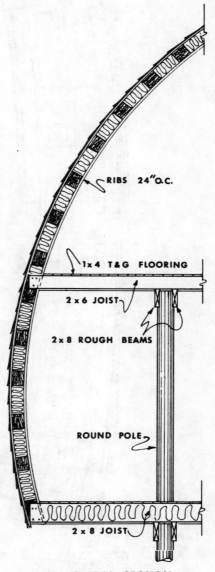

STRUCTURAL SECTION

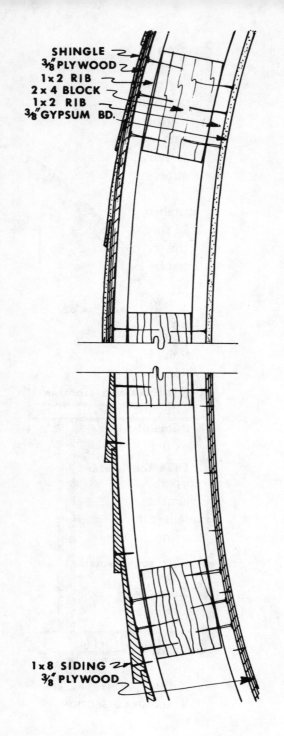

Utility Space and Utilities

An enclosure for utility space is provided under both sections below the first floor. This is an on-grade concrete slab, over which either a concrete block wall, or a conventional wood-frame wall of treated lumber and plywood is erected. This provides space, entered from outside, for the water heater and heating system and for washing and ironing.

Electrical wiring is installed conventionally. Horizontal runs are easy to install through openings between the block web members in the arches. Plumbing and heating is conventional, as required by local codes.

Additional Information

The Housing Research Unit of the Southeastern Forest Experiment Station is not an architectural design agency. It has developed this and other designs primarily to illustrate new and effective ways to use wood and wood products more efficiently in house construction.

A prototype of duplex unit has been built privately in Athens, Georgia. Experience gained in construction of the prototype was incorporated in the revised plans now available. Room arrangements, door and window placement, and interior details can be varied, because the interior walls are nonbearing.

Designed By Harold F. Zornig
Southeastern Forest Experiment Station

Index

Index

A

Access doors	373
Aluminum foil	45, 119
American Softwood Lumber Standard	79

B

Balusters	328
Barns, material	34
Baseboard	299
Bathroom painting	353
Bathtub framing	247
Battens	131, 164, 228
Batt insulation	255
Beams	13, 36, 128, 147
Blanket insulation	132
Boards	83
Boiled linseed oil	332

C

Cabinets	299
Casings, door	287
Ceilings	211
finish	264
tile	133
Center beams,	358
drop	358
flush	357

Chimney 137, 214
Clear film finishes 333
Collar beam 209
Combination doors 131
Commercial Standards 47
Common grade 70
Concrete forms, meterial 42
Condensation 106
Construction grade 73
Cornice 214
Costs 77, 121, 130, 164
Crawl space 122, 155, 373

D

Decay 105
    prevention 105
    resistance 62, 76, 105
Decking. 26
    material 26
    roof 122, 193
Dimensional stability 50, 96
Dimension grade 70, 73, 128
Door 287
    casings 287
    frames 89, 233, 244, 284
    stops 289
Door hardware 290
    hinges 290
    locks 291
    strike plate 292
Doors 31, 32, 131, 133, 235, 285
Dressed dimensions 78
Drop center beams 358
Dry-wall 132
    coverings 132
    painting 353

E

Elevation stakes 357
Exterior 329
    painting 329
    plywood 90
    trim material 23, 87, 130, 203
    walls 123

F

Facia	130, 203
Factory grades	73
Felt	45, 46, 106
Fences	41
gates	41
posts	40, 68, 126, 132, 149
Fiberboard	10
Fill insulation	118, 132
Finish grade, wood	73
Finishing	136
bathroom walls	275
gypsum board	265
hardboard	269
paneling	272
tile	273
Flexible insulation	118
Floor coverings	276
Floor system	122, 128
Flooring	28, 86, 122
strip	277
tile	282
wood block	282
Flooring material	28
block	29, 86, 95
strip	29
underlay	28
Flush center beams	356
Footings	356
Foundation	13, 35, 139
Foundation system	139
beams	147
chemical	153, 155
edge piers	149
footings	142
insulation	164
joists	156
ledgers	150
posts	142, 149
site selection	139
termite protection	153
Framed wall system	171
end	171
interior	179
side	165
walls	165

Framing 15
    grade 79
    headers 15
    joists 15
    plates 16
    rafters 15
    studs 16, 128
Frames 89
    door 89
    window 89

G

Gable end 214
Gloss paint 335
Grade 69, 71
Grain bin 73
Gusset trusses 205
Gypsum 132
    board 265

H

Handrail 328
Hardness, wood 49
Hardwood 12, 48, 73
    price 77, 121, 130
    odor 65
Hardwood grades 73
    construction 73
    dimension 73
Headers 15
Heartwood proportion 63
Hinges 290
High-density hardboard 93
House plans 375
    curved walls 410
    hillside duplex 425
    round 419
    24 by 24 ft 385
    24 by 32 ft 378
    24 by 32 ft expandable 391
    24 by 28 ft 393
    24 by 36 ft 381
    28 by 36 ft 401
    28' 4" by 36' 4" 405
    32 by 32 ft 397

I

Inlet ventilators	259
Insulation	117
considerations	117, 132, 164
materials	118, 132, 164
Interior	334
painting	334
plywood	90
trim, material	26, 94
walls	123

J

Jambs	131, 133
Joint	265
cement	265
tape	267
Joists	15, 37, 128, 156

K

Kitchen	301
arrangements	301
painting	353
cabinets	301
base	302
wall	303

L

Laminated paper	254
Latex paint	335
Locks	291
Loose-fill insulation	256
Low-slope roof	184

M

Masonry foundations	354
Materials, barns	35
foundations	35
feed racks	40
joists	37
rafters	37
roof	37, 45
siding	38, 129
Materials, house	13
beams	13, 14, 36, 122, 128
exterior trim	23, 130

  foundation 13
  plates 14
  plank roof decking 21, 122
  shakes 22, 132
  shingles 22, 88
  sills 13
  sleepers 14
  subfloors 16
  wall sheathing 20, 45, 83, 129, 165
Medium hardboard 10, 92
Metal plate gussets 205
Millwork 94, 133
Molding 87
Moisture content 95
  protecting 99, 105
  testing 96
  upon installation 96
Moisture meter 96

N

Nails 136
  holding ability 59
National 248
  product standards 47

O

Oil-based preservatives 113
Outdoor steps, material 26
Outlet ventilators 258

P

Painting 114, 136, 328
  bathroom 353
  dry wall 353
  exterior 329
  interior 334
  kitchen 353
  plaster 353
  trim 335
Paint performance (of wood) 58
Paneling 33
Panel siding 226
Paper 45, 46, 117
Particleboard 10, 93
Piers 122

Pitched roof	203
Plank roof decking	21, 37
Plaster painting	353
Plates	14, 16, 36
Platforms, materials	44
Plumbing	246
Plywood	10, 89, 269
Poles	34, 126
Polyethylene	353
Porches	307
construction	310
floor	310
gable roof	312
low-slope roof	312
materials	44
Preservatives	111, 126
applying	113
oils	113
salts	113

R

Rafter-joist roof	189
Rafters	15, 37
Reflective insulation	119
Regular hardboard	10
Ridge board	208
insulation	118
Ring connectors	205
Roof construction	123, 126
framing	123
insulation	196
trim	203, 213
trusses	123
Roof coverings	218
asphalt shingles	218
Boston ridge	220
Roof systems	179
general	179
insulation	196
low-slope	184
pitched roof	203
rafter-joist	189
sheathing	211
trim	203, 213
trussed roof	205
wood deck	193
Rool roofing	45, 130

S

Saline preservatives 113
Scab 195
Scaffolding, materials 43
Sealer 277
Select grade, wood 60, 70, 73
Semigloss paint 335
Shakes, material 22
Sheathing 25, 83, 129, 187, 211
Sheet materials 45
    aluminum foil 45, 119
    felt 45, 106
    paper 45, 117
    roofing 45
Shelving 32
Shingles 22, 80, 130, 188, 218
Shrinkage 50, 96
Siding 24, 45, 84, 129, 230
    bevel 231
    vertical 232
Sills 13, 36, 131
Skirtboard 367
    foundation 367
    house 367
Softwood 12, 48, 71, 73
Soil cover 353, 372
Spackle compound 267
Span 128
Stains 331
Stairs 321
    inside 321
    outside 317
Stairways 32
Steps 317
    foundation 364
Stool and apron 293
Stoop 319
Stops 131
Storage bins, materials 44, 73
Strike plate 292
Strip flooring 277
Structural 10
    insulating board 10, 91, 118, 132, 226
    properties 47
Subfloors 16, 122

T

Tanks, materials	44
Thermal insulation	118, 132, 251
Thinners	333
Tile	282
ceiling	133
Trussed roof	205
Trusses	205
gusset	205
metal plates	205
ring connectors	205

U

Unfinished wood	329
Utilities	137
U value	198

V

Vapor barriers	45, 118, 119, 251, 253
Vats, materials	44
Vermiculite	252
Ventilation	106, 373
attic	258
crawl space	261
inlet ventilators	259
outlet ventilators	258

W

Walls,	123
exterior	123
finish	264
interior	123
sheathing	20, 45, 129, 131, 165
Wardrobe-closet combinations	307
Water-repellent finishes	330
Wear resistance	76
Weight	50
Windows	235
casings	292
frames	89, 233, 242
stops	298
Wiring	248
Wood-post foundations	353
Woods, grades	69, 71
common	70

    dimension 70, 73, 79
Woods, structural properties 47
    bending strength 65
    compression strength 68
    decay resistance 62, 76, 105
    dimensional stability 50, 96
    figure 64
    hardness 49
    hardwoods 48, 73
    heartwood proportion 63
    nail holding 59
    paint performance 58
    softwoods 48, 73
    stiffness 67
    toughness 69
    weight 50
    workability 53
Workability (of wood) 53

Y

Yard lumber 73

NOTES

NOTES

NOTES